Lecture Notes in Mathematics 1498

Editors:
A. Dold, Heidelberg
B. Eckmann, Zürich
F. Takens, Groningen

Reinhard Lang

Spectral Theory of Random Schrödinger Operators

A Genetic Introduction

Springer-Verlag
Berlin Heidelberg New York
London Paris Tokyo
Hong Kong Barcelona
Budapest

Author

Reinhard Lang
Institut für Angewandte Mathematik
Im Neuenheimer Feld 294
6900 Heidelberg, Germany

Mathematics Subject Classification (1991): 60-02, 60H25, 35P05, 35R60, 81C20, 82A42

ISBN 3-540-54975-7 Springer-Verlag Berlin Heidelberg New York
ISBN 0-387-54975-7 Springer-Verlag New York Berlin Heidelberg

Printed in Germany

Typesetting: Camera ready by author
Printing and binding: Druckhaus Beltz, Hemsbach/Bergstr.
46/3140-543210 - Printed on acid-free paper

Dedicated to Frank Spitzer on the occasion of his sixty fifth birthday.

Preface

These notes are taken from seminars on the spectral theory of random Schrödinger operators, held at the University of Heidelberg during 1988 and 1989. Addressed to the non-specialist they are intended to provide a brief and elementary introduction to some branches of this field. An attempt is made to show some of the basic ideas *in statu nascendi*, and to follow their evolution from simple beginnings to more advanced results. The term "genetic" in the title refers to this procedure.

The main theme is the interplay between the spectral theory of Schrödinger operators and probabilistic considerations. After developing a general intuitive picture to give the reader some orientation in the field, we elaborate on two topics which in the history of the subject have proved to be of major conceptual importance. We consider on the one hand the Laplacian in a random medium and study its spectrum near the left end, where large fluctuations in the medium play an essential role. Expressed in terms of Wiener measure, this amounts to large deviation problems for Brownian motion. Guided by these questions we show how the notion of *entropy* has undergone mutations, and explain its relation to the spectral theory of the Laplacian. On the other hand we specialize to one-dimensional space and consider there Schrödinger operators with general ergodic potentials. We explain how certain aspects of the Floquet theory can be extended from periodic to general potentials. Based on this extension, the absolutely continuous spectrum of one-dimensional Schrödinger operators is studied. Here the notion of *rotation number* and its relation to Weyl's theory of singular Sturm-Liouville operators play an important role.

An effort is made throughout to give heuristic arguments before going to rigorous proofs. By means of a few characteristic problems and their solution we attempt to explain basic ideas and concepts in the simplest possible setting rather than to collect the most refined results.

These notes are dedicated by a grateful disciple to Frank Spitzer on the occasion of his sixty fifth birthday.

Table of contents

Structure of the notes

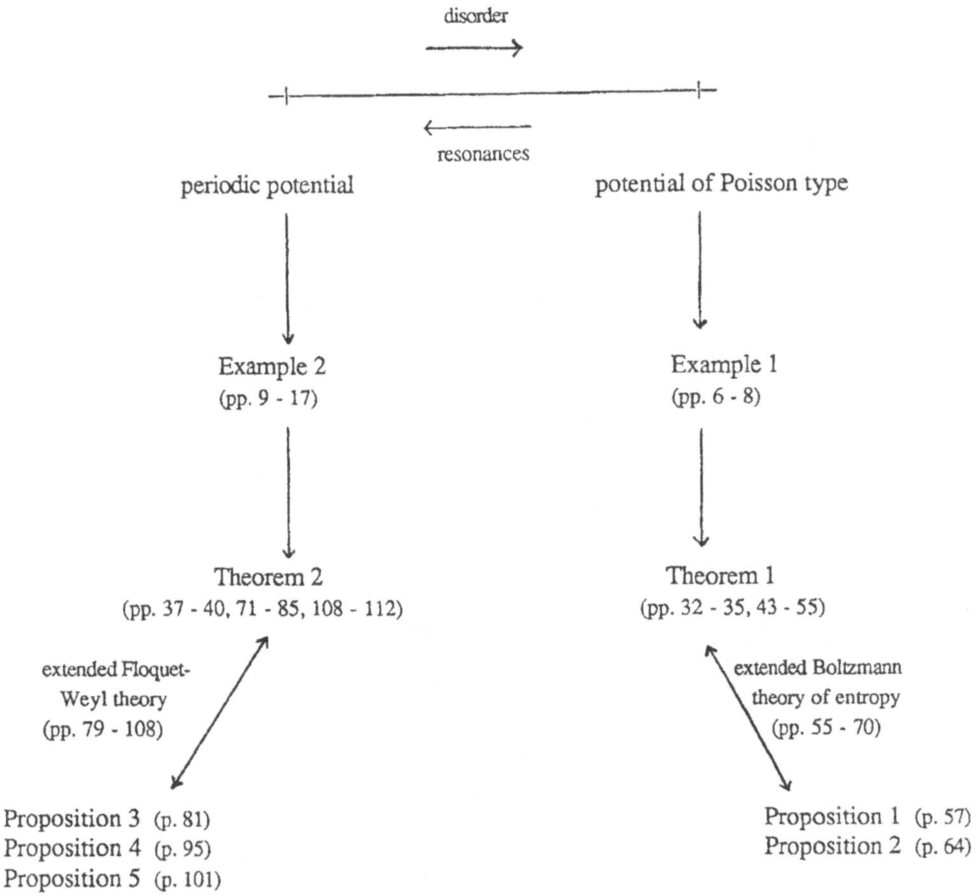

Introduction

We consider Schrödinger operators

.1) $H = -\Delta + q,$

where $\Delta = \frac{\partial^2}{\partial x_1^2} + \ldots + \frac{\partial^2}{\partial x_d^2}$ denotes the Laplacian in $\mathbf{R}^d$, $d \geq 1$, and $q: \mathbf{R}^d \to (-\infty, +\infty]$ is a random potential, and address ourselves to the following <u>general problem</u>:

.2) What is the spectrum $\Sigma = \Sigma(q)$ of the operator $H = H(q)$ and what do the solutions g_λ to the equation $Hg_\lambda = \lambda \cdot g_\lambda$, $\lambda \in \Sigma$, look like? How does the answer depend on the degree of disorder of the potential q?

A solution g_λ to the equation

.3) $Hg_\lambda = \lambda \cdot g_\lambda$

is called an eigenfunction corresponding to the eigenvalue λ, if $g_\lambda \in L^2(\mathbf{R}^d)$. The spectrum Σ is the union of the pure point spectrum Σ_{pp}, which is defined as the closure of the set of eigenvalues, and of the continuous spectrum Σ_c, the precise definition of which is postponed to the end of this section; roughly speaking, λ belongs to Σ_c, if equation (1.3) has approximate solutions in $L^2(\mathbf{R}^d)$ which are orthogonal to the space spanned by the eigenfunctions. As will be seen later, this notion has still to be refined by splitting Σ_c into an absolutely continuous part and a singularly continuous part.

Problem (1.2) has at least two quite different origins coming from (i) solid state physics and (ii) purely mathematical considerations.

(i) In order to see how it arises in solid state physics, we consider a solid whose atoms are located at sites $x_i \in \mathbf{R}^d$, $i \geq 1$, and a pair potential

.4) $\Phi: \mathbf{R}^d \to (-\infty, +\infty],$

describing the interaction between a particle and an atom. In the one-body approximation one considers the Schrödinger operator (1.1) with potential

$$(1.5)\quad q(x) = \sum_i \Phi(x - x_i) \ , \ \ x \in \mathbf{R}^d .$$

In the case of a perfect crystal, the x_i are points of a lattice, e.g. of the lattice $\mathbf{Z}^d$, so that q is a deterministic periodic function. In general there are random deviations from the lattice structure and the points x_i are randomly distributed. One is in a case of maximal disorder if the x_i are distributed according to a Poisson point process on $\mathbf{R}^d$. The degree of randomness may range from periodic structures (respectively periodic potentials q) to structures with strong randomness (respectively to strongly mixing potentials q).

degree of randomness

→

perfect crystal	Poisson points
periodic potential	potential of Poisson type (1.5)

Figure 1

The disorder gives rise to qualitatively new properties of the solid state which cannot be understood merely on the basis of perturbations around the case of perfect order. It is a fundamental problem to understand the conditions on the potential q under which the solutions g_λ of (1.3) are exponentially decreasing at infinity (localization) or wavelike (extended states). Physically, the transition from extended states to localized states corresponds to the transition from a metal to an insulator (see Anderson (1958)).

(ii) On the mathematical side one tries to understand the spectral properties of the operator (1.1) for a given deterministic potential q. Until the seventies mathematical results existed for three classes of potentials. If $q(x) \to +\infty$ as $|x| \to \infty$, H has purely -

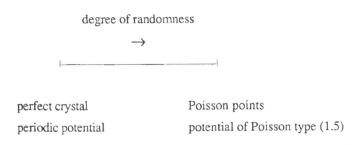

discrete spectrum. A characteristic example is the case of a vibrating membrane D with fixed boundary, which formally corresponds to the potential

$$q(x) = \begin{cases} 0, & x \in D \\ +\infty, & x \notin D \end{cases},$$

where D denotes a bounded open set in $\mathbf{R}^d$. The second class is the class of potentials rapidly decreasing at infinity, which is treated by scattering theory. A characteristic example is given by a non-negative potential with compact support where the spectrum $\Sigma = [0,\infty)$ is purely continuous and where one has wavelike solutions g_λ for $\lambda \in \Sigma$. Thirdly, in the case of a periodic potential, one has purely continuous spectrum and Bloch waves as eigenstates. In all other cases however, if q is bounded and oscillating but not strictly periodic, very little is known about the spectral properties of H. Since it is difficult to get results for an individual non-periodic bounded q, one randomizes the problem and contents oneself with asking what the spectral behaviour of H is for typical bounded random potentials. A similar probabilistic approach was introduced by Bloch and Pólya (1932) in order to get estimates about the number of real roots of a real polynomial of high degree. This problem serves Kac as a striking example illustrating the nature and power of probabilistic reasoning, see p. 11 in Kac (1959).

From these remarks on the different origins of problem (1.2) it is already clear that it has connections with many fields, ranging, as we will see, from Statistical Mechanics to the theory of integrable Hamiltonian systems. This diversity of the subject makes its beauty but also its difficulty. There exists a rich literature, for example a monograph of encyclopedic character by Carmona and Lacroix (1990), and also several expositions and review papers. A sample of the more recent ones are Spencer (1986), Cycon, Froese, Kirsch and Simon (1987), Martinelli and Scoppola (1987), Bellissard (1989), Pastur (1989). It is the aim of the present paper, to give an elementary introduction to some basic problems and ideas in the field outlined above and to follow their evolution approximately in historical order. The paper assumes no particular background. It is addressed to the general reader and mainly deals with the following questions.

(i) What is the heuristic picture underlying problem (1.2)? What kind of mathematical questions arise from this picture? What are the relations between its several aspects?

(ii) How did ideas evolve from the simple beginnings of spectral theory to more advanced results related to the clarification of (1.2)? How did, conversely, probabilistic considerations lead to a mutation and advancement of classical theories?

Sections 2 - 4 are devoted to question (i). Here a general heuristic picture is drawn which suggests, roughly speaking, a tendency to localization with increasing disorder. In sections 5 and 6 we deal with question (ii) by way of two characteristic problems. Along with their solution, we try to explain how the classical notions of entropy and of Floquet exponent were extended to general concepts of wide applicability.

In section 2 we begin with two quite elementary examples. The first, which is located at the right end point on the scale of disorder in Figure 1, deals with one-dimensional potentials of Poisson type. The second example is located at the left end point on the scale in Figure 1 and concerns one-dimensional periodic potentials. Guided by these examples we sketch in section 3 some heuristic ideas concerning localization and the asymptotic behaviour of the density of states near to the bottom of the spectrum (so called Lifschitz tails). Section 4 contains a discussion about the present mathematical status of these ideas and some of the main open problems are mentioned. In order to illustrate question (ii), we have chosen two key results by Donsker and Varadhan (1975a,b,c) and by Kotani (1984) respectively, whose proofs are conceptually significant and basic for many other work too, and we try to explain their contents in detail in sections 5 and 6. Theorem 1, which can be seen as generalization of example 1 to higher dimensions, deals with Lifschitz tails in a Poisson model. Along with a sketch of its proof in section 5, an extension of the Boltzmann theory of entropy is given. Theorem 2 is inspired by example 2. It says, roughly speaking, that randomness implies the absence of the absolutely continuous spectrum, if the dimension is one. Its proof requires a far reaching extension of the Floquet theory as will be explained in section 6. Section 5 and 6 are formally independent of each other. Parallel reading however could help the reader to get a balanced picture of probabilistic and deterministic aspects of the theory: on the random side of Figure 1 large fluctuations in the medium and correspondingly the theory of large deviations are relevant, whereas on the deterministic side conserved quantities and the notion of a generalized Floquet exponent play an essential role. In the final section we reflect the meaning of what has been done in the previous sections and we reconsider the development from the simple vibrating string to the spectral theory of infinitely many randomly coupled vibrating membranes.

Before we begin with the discussion of the two examples let us give some definitions. We denote by $L^2(\mathbf{R}^d)$ the space of measurable functions u: $\mathbf{R}^d \to \mathbf{C}$ with finite norm $\|u\| = (\int |u(x)|^2\, dx)^{1/2}$ and by $\mathcal{H}_{pp}$ the subspace of $L^2(\mathbf{R}^d)$ spanned by the eigenfunctions of H. The continuous spectrum Σ_c is defined as follows. A number $\lambda \in \mathbf{R}$ belongs to Σ_c if and only if there exists a sequence $u_n \in L^2(\mathbf{R}^d)$ such that u_n is orthogonal to $\mathcal{H}_{pp}$, $\|u_n\| = 1$ for all $n \in \mathbf{N}$, and $\|(H - \lambda)u_n\| \to 0$ as $n \to \infty$.

The following general probabilistic framework is used. We denote by Ω a subspace (which has to be specified according to the context in which we are working) of the space of measurable functions q: $\mathbf{R}^d \to (-\infty, +\infty]$ and by $(\Omega, \mathcal{F}, P)$ a probability space. The shift operator θ_x: $\Omega \to \Omega$ is defined by $\theta_x q(y) = q(x + y)$ for $x,y \in \mathbf{R}^d$.

We assume that P satisfies

.6) $$P(\theta_x A) = P(A) \quad \text{for } A \in \mathcal{F},\ x \in \mathbf{R}^d \text{ (shift invariance)}$$

and

.7) $$\text{if } A \in \mathcal{F} \text{ and } P((\theta_x A)\,\Delta\, A) = 0 \text{ for all } x \in \mathbf{R}^d, \text{ then } P(A) = 0 \text{ or } P(A) = 1 \quad \text{(ergodicity).}$$

The ergodicity of P implies that many quantities are selfaveraging, see for example (2.6) below. We say a property holds almost everywhere with respect to P (abbreviated by P-a.e.), if the set of potentials with this property has P-measure 1.

In the following we do not enter into questions about the selfadjointness of the operator H or into measurability questions, and refer instead to the literature, e.g. to Carmona and Lacroix (1990).

2. Two simple examples

In order to get a feeling for problem (1.2) we begin with two quite elementary one-dimensional examples, which are chosen as extreme cases on the scale of disorder in Figure 1.

2.1. Example 1

In the first example we consider the one-dimensional potential of Poisson type

$$q(x) = \sum_{i \in \mathbf{Z}} \Phi(x - x_i) , \tag{2.1}$$

where x_i, $i \in \mathbf{Z}$, are the points of a Poisson process on $\mathbf{R}^1$, say with average density 1, and where

$$\Phi(x) = \begin{cases} +\infty , & |x| \leq R \\ 0 , & |x| > R \end{cases} \tag{2.2}$$

is a hard core potential of a given radius $R > 0$. The corresponding operator $H = H(q)$ is just the operator $-\frac{d^2}{dx^2}$ in the random region $\mathbf{R} \setminus \bigcup_{i \in \mathbf{Z}} [x_i - R, x_i + R] = \bigcup_{i \in \mathbf{Z}} D_i$ with Dirichlet boundary conditions, see Figure 2.

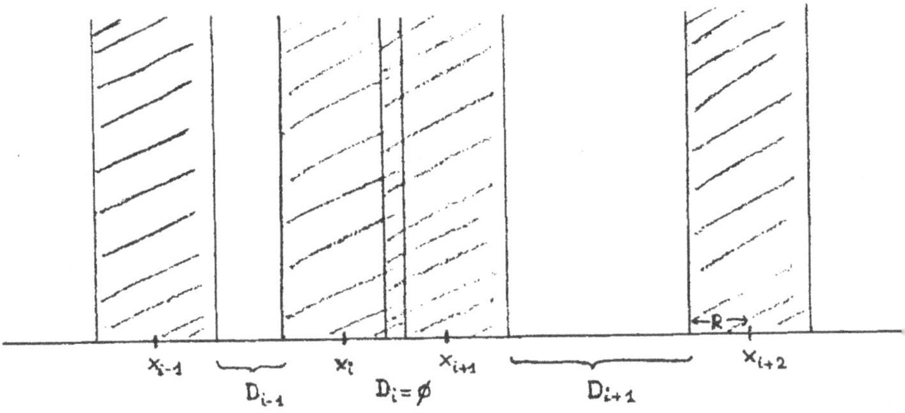

Figure 2

The spectral behaviour of H is the following:

(2.3) $\Sigma = \Sigma_{pp} = [0,\infty)$, the eigenvalues are P – a.e. dense in Σ

and

(2.4) the eigenfunctions have compact supports P – a.e..

This is obvious since the D_i, $i \in \mathbf{Z}$, can be viewed as strings of random length $|D_i|$ which vibrate *independently* of each other. The eigenvalues corresponding to a string D_i with $D_i \neq \emptyset$ are given by

$$\lambda_{i,j} = \frac{\pi^2 \cdot j^2}{|D_i|^2}, \qquad j \geq 1 . \tag{2.5}$$

The numbers $\{ |D_i|, i \in \mathbf{Z} \}$ and therefore the eigenvalues $\{\lambda_{i,j}, (i,j) \in \mathbf{Z} \times \mathbf{N}$ with $D_i \neq \emptyset \}$ are P – a.e. dense in $[0,\infty)$ and the corresponding eigenfunctions are localized in the intervals D_i. This shows (2.3) and (2.4).

The solutions g_λ to (1.3) and the eigenvalues are random. For general ergodic potentials they are no longer computable explicitly. It is therefore appropriate to introduce another notion, which is defined via an averaging procedure, and hence non-random. This so-called integrated density of states N gives information about "the number of eigenvalues per unit interval", and is defined as follows. For $L > 0$ we denote by $H^{(L)}$ the operator $-\frac{d^2}{dx^2} + q$ in the interval $(-L, +L)$ with Dirichlet boundary conditions. Then

$$N = \lim_{L \to \infty} N^{(L)} \tag{2.6}$$

is the limit of the empirical distributions

$$N^{(L)}(\lambda) = \frac{1}{2L} \# \{ i \in \mathbf{N} : \lambda_i^{(L)} < \lambda \}, \quad \lambda \in \mathbf{R}, \tag{2.7}$$

where $\lambda_i^{(L)}$, $i \geq 1$, are the eigenvalues of the operator $H^{(L)}$. In the case of the Poisson potential (2.1), it is easy to show the existence of the limit (2.6) and to compute, for $\lambda \geq 0$,

(2.8) $$N(\lambda) = \lim_{L\to\infty} \frac{1}{2L} \#\{ (i,j)\in \mathbf{Z} \times \mathbf{N}: D_i \subset (-L,+L),\ j \le \frac{\sqrt{\lambda}}{\pi} |D_i| \} =$$

$$= E[\frac{1}{\pi} \sqrt{\lambda}\, |D_1|\,],$$

where [x] denotes the greatest integer which is $\le x$. In particular, one gets from (2.8) the following asymptotic behaviour of N near the boundaries of the spectrum:

(2.9) $$N(\lambda) \sim \begin{cases} e^{-\pi/\sqrt{\lambda}} & , \lambda \to 0 \\ e^{-2R} \frac{1}{\pi} \sqrt{\lambda} & , \lambda \to \infty , \end{cases}$$

which more precisely means

(2.10) $$\lim_{\lambda\to 0} \sqrt{\lambda} \log N(\lambda) = -\pi ,$$

respectively

(2.11) $$\lim_{\lambda\to\infty} \pi\, e^{2R} \frac{1}{\sqrt{\lambda}} N(\lambda) = 1 .$$

The asymptotic behaviour of N as $\lambda \to \infty$ is, up to a constant factor, clearly the same as in the free case $q \equiv 0$, where the integrated density of states is $N_0(\lambda) = \frac{1}{\pi} (\max\{0,\lambda\})^{1/2}$. To see the probabilistic meaning of the asymptotic behaviour of N as $\lambda \to 0$ we denote by D(r) the open ball of radius r, which in the present one-dimensional case is just an interval of length 2r, and to a given $\lambda > 0$ we determine the radius $r(\lambda) = \frac{\pi}{2\sqrt{\lambda}}$ so that the lowest Dirichlet eigenvalue of the operator $-\frac{d^2}{dx^2}$ in the interval $D(r(\lambda))$ is λ. Then we have

(2.12) $$N(\lambda) \sim e^{-\pi/\sqrt{\lambda}} = P(D(r(\lambda)) \text{ is free of Poisson points}),\ \lambda \to 0.$$

That is, near to the bottom of the spectrum, the essential contribution to $N(\lambda)$ comes from large intervals D_i which have approximately λ as lowest Dirichlet eigenvalue and which are free of Poisson points (see Figure 2 above).

2.2. Example 2

In the second example we consider a potential which is a continuous periodic function $q: \mathbf{R}^1 \to \mathbf{R}^1$, say with period L. Formally this case can be subsumed under the general framework indicated at the end of the introduction as follows. One chooses a point ω equidistributed in the interval [0,L] and deals with the random potential $\theta_\omega q$ instead of q. For the purpose of this section such a randomization is not needed, and we consider here the deterministic operator

.13)
$$H = -\frac{d^2}{dx^2} + q \,, \qquad -\infty < x < +\infty \,,$$

and the corresponding equation

.14)
$$H g_\lambda = \lambda \cdot g_\lambda \,, \qquad \lambda \in \mathbf{C} \,.$$

The following result holds true. There exist real numbers a_i, $i \geq 0$, with $-\infty < a_0 < a_1 \leq a_2 < a_3 \leq a_4 < \dots$, such that

.15)
$$\Sigma = \Sigma_c = \bigcup_{i \geq 0} [a_{2i}, a_{2i+1}] \,,$$

and for $\lambda \in \bigcup_{i \geq 0} (a_{2i}, a_{2i+1})$ there exist two linearly independent solutions of (2.14) of the form

.16)
$$g_\pm(\lambda; x) = p_\pm(\lambda; x)\, e^{\pm i \alpha(\lambda) x} \,, \quad x \in \mathbf{R},$$

where $p_\pm(\lambda; \cdot)$ is a complex valued function with period L and the phase $\alpha(\lambda)$ is a real number. The intervals $[a_{2i}, a_{2i+1}]$ are the allowed energy bands and the (a_{2i-1}, a_{2i}) are (possibly empty) gaps in the spectrum; the solutions (2.16) are called Bloch waves.

We indicate the four main steps of the proof of (2.15) and (2.16) with a view towards extensions in later sections. Details can be found for example in the booklet by Magnus and Winkler (1979).

Step 1. Transformation of equation (2.14) into a dynamical system.

This step does not rely on the periodicity of q, it can be performed for arbitrary one-dimensional potentials. One can write (2.14) in the form

(2.17)
$$\begin{pmatrix} g_\lambda(x) \\ g_\lambda'(x) \end{pmatrix}' = \begin{pmatrix} 0 & 1 \\ q(x)-\lambda & 0 \end{pmatrix} \begin{pmatrix} g_\lambda(x) \\ g_\lambda'(x) \end{pmatrix},$$

with $g_\lambda'(x) = \frac{d}{dx} g_\lambda(x)$. We denote by $\begin{pmatrix} \varphi_\lambda(x) \\ \varphi_\lambda'(x) \end{pmatrix}$ and $\begin{pmatrix} \psi_\lambda(x) \\ \psi_\lambda'(x) \end{pmatrix}$ the solutions of (2.17) with $\begin{pmatrix} \varphi_\lambda(0) \\ \varphi_\lambda'(0) \end{pmatrix} = \begin{pmatrix} 1 \\ 0 \end{pmatrix}$ and $\begin{pmatrix} \psi_\lambda(0) \\ \psi_\lambda'(0) \end{pmatrix} = \begin{pmatrix} 0 \\ 1 \end{pmatrix}$ respectively, and by

(2.18)
$$Y_\lambda(x) = Y_\lambda(x;q) = \begin{pmatrix} \varphi_\lambda(x) & \psi_\lambda(x) \\ \varphi_\lambda'(x) & \psi_\lambda'(x) \end{pmatrix}$$

the fundamental matrix of equation (2.17). The following conservation law is basic and has important consequences as we will later see. The Wronski determinant

(2.19)
$$[f_\lambda, g_\lambda]\ (x) = \det \begin{pmatrix} f_\lambda(x) & g_\lambda(x) \\ f_\lambda'(x) & g_\lambda'(x) \end{pmatrix}$$

of two solutions f_λ, g_λ of (2.17) is constant. This can easily be seen by differentiation of the determinant. In particular one has

(2.20)
$$\det Y_\lambda(x) = [\varphi_\lambda, \psi_\lambda]\ (x) \equiv 1\,, \quad x \in \mathbf{R}.$$

Step 2. Floquet theory.

In order to find out for which $\lambda \in \mathbf{C}$ equation (2.14) has wavelike solutions, one tries to understand the stability behaviour of the dynamical system (2.17). To this end one takes advantage of the periodicity of q and asks for solutions g_λ which, after turning one period, only change by a (in general complex) factor, i.e. for solutions g_λ and factors $\mu(\lambda) \in \mathbf{C}$ such that

(2.21)
$$g_\lambda(x+L) = \mu(\lambda) \cdot g_\lambda(x)\,, \quad x \in \mathbf{R}.$$

Such solutions do indeed exist and the multipliers can be written as

(2.22) $$\mu_{\pm}(\lambda) = e^{\pm w(\lambda)\cdot L},$$

where $w(\lambda) \in \mathbf{C}$ is the so called Floquet exponent corresponding to equation (2.14). For the proof one sets $g_\lambda = c \cdot \varphi_\lambda + d \cdot \psi_\lambda$ with coefficients $c, d \in \mathbf{C}$ still to be determined. From (2.21) one gets

$$Y_\lambda(x+L)\binom{c}{d} = \mu(\lambda) \cdot Y_\lambda(x)\binom{c}{d}, \qquad x \in \mathbf{R}.$$

With $x = 0$ it follows that $\mu(\lambda)$ is one of the two eigenvalues $\mu_{\pm}(\lambda)$ of $Y_\lambda(L)$, which because of (2.20) can be written in the form (2.22). Using the special solutions with property (2.21) one can furthermore show, that for all $\lambda \in \mathbf{C}$ there exist two *linearly independent* solutions of (2.14) which are of the form

(2.23) $$g_{\pm}(\lambda;x) = p_{\pm}(\lambda;x)e^{\pm w(\lambda)\cdot x} \qquad , \text{ if } iw(\lambda) \notin \frac{\pi}{L}\cdot \mathbf{Z},$$

(2.24) $$\left\{\begin{aligned} g_+(\lambda;x) &= (p_+(\lambda;x)) + \varepsilon\cdot x\cdot p_-(\lambda;x))\, e^{+w(\lambda)\cdot x} \\ g_-(\lambda;x) &= p_-(\lambda;x)\, e^{+w(\lambda)\cdot x} \end{aligned}\right\} \qquad , \text{ if } iw(\lambda) \in \frac{\pi}{L}\cdot \mathbf{Z},$$

where $p_{\pm}(\lambda;\cdot)$ is a complex valued function with period L and $\varepsilon \in \{0,1\}$. The solutions $g_{\pm}$ are called Floquet solutions.

Step 3. Stability and spectrum.

In this and in the following step let λ be real. We split w into its real and imaginary part

(2.25) $$w(\lambda) = -\gamma(\lambda) + i\alpha(\lambda), \qquad \lambda \in \mathbf{R},$$

assuming $\gamma(\lambda) \geq 0$ and $\alpha(\lambda) \geq 0$ by convention. $\gamma(\lambda)$ describes the change of amplitude of the Floquet solutions as x is increased by a period L and $\alpha(\lambda)$ describes the corresponding change of phase.

The spectrum $\Sigma(q)$ and the stability behaviour of the dynamical system (2.17) are related as follows:

(2.26) $$\Sigma = \Sigma_c = \{\lambda \in \mathbf{R}:\ \gamma(\lambda) = 0\}.$$

This can easily be seen with the help of the Floquet theory described in step 2. For the proof we split proposition (2.26) into three parts.

(2.27) $$\Sigma_{pp} = \emptyset$$

(2.28) $$\gamma(\lambda) = 0 \ \Rightarrow \lambda \in \Sigma$$

(2.29) $$\gamma(\lambda) > 0 \ \Rightarrow \lambda \notin \Sigma .$$

(2.27) is clear since none of the solutions listed in (2.23) and in (2.24) is in $L^2(\mathbf{R}^1)$. According to this list, a bounded solution exists in the case $\gamma(\lambda) = 0$. Multiplying such a solution with a function, which is constant for $|x| \leq n$ and vanishing for $|x| \geq n+1$, one can find approximating functions $u_n \in L^2(\mathbf{R}^1)$ such that $\|u_n\| = 1$ $(n \geq 1)$ and $\|(H - \lambda)u_n\| \to 0$ as $n \to \infty$. This implies $\lambda \in \Sigma_c = \Sigma$. In view of later sections the proof of (2.29) is of some interest since it clarifies the relation between the spectrum and stability properties of the dynamical system (2.17). If $\gamma(\lambda) > 0$, one is in the case (2.23). Then there exist two linearly independent solutions $g_\pm(\lambda, \cdot)$ such that

(2.30) $$|\, g_\pm(\lambda;x)\,| \leq \text{const.}\ \ e^{\mp\gamma(\lambda)\cdot x} \ , \ -\infty < x < +\infty .$$

Since $g_+(\lambda, \cdot)$ and $g_-(\lambda, \cdot)$ are linearly independent, one obtains from (2.19)

$$[g_+(\lambda), g_-(\lambda)]\,(x) \equiv [g_+(\lambda), g_-(\lambda)](0) \neq 0.$$

Hence,

(2.31) $$G_\lambda(x,y) = G_\lambda(y,x) = \frac{g_-(\lambda;x)\cdot g_+(\lambda;y)}{\left[\, g_+(\lambda)\,,\, g_-(\lambda)\,\right]} \ , \quad x \leq y \ ,$$

is well defined and one easily verifies

(2.32) $$(H - \lambda)\ G_\lambda(x,y) = \delta(x - y) \ ,$$

i.e. G_λ is just the Green's function of the operator $H - \lambda$. From (2.30) one obtains (the proof is postponed to the end of section 2.2.), that

$$f \mapsto G_\lambda f(x) = \int G_\lambda(x,y)\, f(y)\, dy$$

defines a bounded operator on $L^2(\mathbf{R}^1)$. Hence $(H-\lambda)^{-1}$ exists as a bounded operator on $L^2(\mathbf{R}^1)$. Therefore, according to the definition given at the end of the introduction, λ does not belong to the continuous spectrum, since any sequence $u_n \in L^2(\mathbf{R}^1)$ with $\| (H-\lambda)u_n \| \to 0$ converges to zero because of

$$\|u_n\| = \|(H-\lambda)^{-1} (H-\lambda)u_n \| \le \|(H-\lambda)^{-1}\| \; \|(H-\lambda)u_n \| \to 0 \, .$$

This concludes the proof of relation (2.26).

Step 4. Discriminant and intervals of stability.

In order to understand the band structure of the spectrum, we have to look more carefully at the matrix $Y_\lambda(L)$. We denote its trace, the so called discriminant, by $\Delta(\lambda) = \operatorname{trace} Y_\lambda(L)$. Since $\det Y_\lambda(L) = 1$, the equation for the multiplier $\mu(\lambda)$ is

$$(\mu(\lambda))^2 - \Delta(\lambda)\cdot\mu(\lambda) + 1 = 0$$

with solutions

3) $$\mu(\lambda) = \frac{\Delta(\lambda)}{2} \pm \sqrt{\left(\frac{\Delta(\lambda)}{2}\right)^2 - 1}\, .$$

For real λ, $\Delta(\lambda)$ is real and because of (2.33) one of the following three cases holds:

4) $$|\Delta(\lambda)| > 2\,, \text{ hence } \mu(\lambda)\in\mathbf{R} \text{ and } \gamma(\lambda) > 0$$

5) $$|\Delta(\lambda)| = 2\,, \text{ hence } \mu(\lambda)\in\{\pm 1\}\,,\ \gamma(\lambda) = 0 \text{ and } \alpha(\lambda) \in \frac{\pi}{L}\cdot\mathbf{Z}$$

6) $$|\Delta(\lambda)| < 2\,, \text{ hence } |\mu(\lambda)| = 1\,,\ \gamma(\lambda) = 0 \text{ and } \alpha(\lambda) \notin \frac{\pi}{L}\cdot\mathbf{Z}\, .$$

Because of (2.23) and (2.24) the dynamical system (2.17) is therefore unstable, if (2.34) holds, and it is stable, if (2.36) holds. (2.35) is a boundary case in which one has stability or instability according to the value of ε in (2.24) . As is further shown in Magnus and Winkler (1979), the graph of the real function $\lambda \mapsto \Delta(\lambda)$ looks as in Figure 3.

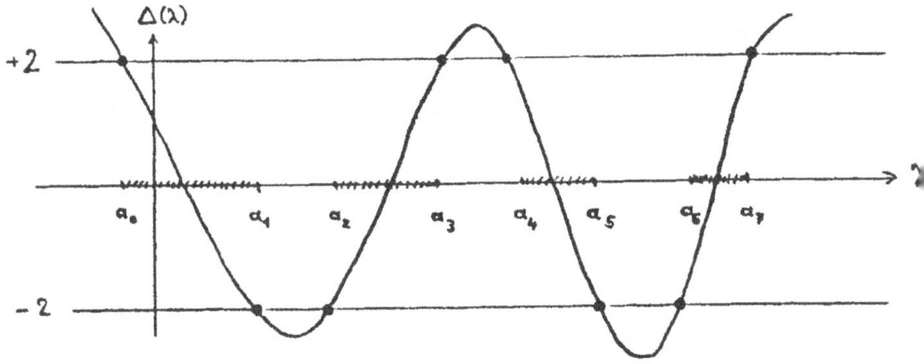

Figure 3

If one keeps the period L fixed and takes the limit $q \to 0$ (e.g. by putting $q(x) = \varepsilon \cdot \cos(2\pi \frac{x}{L})$ and letting $\varepsilon \to 0$), one gets in the limit the following picture:

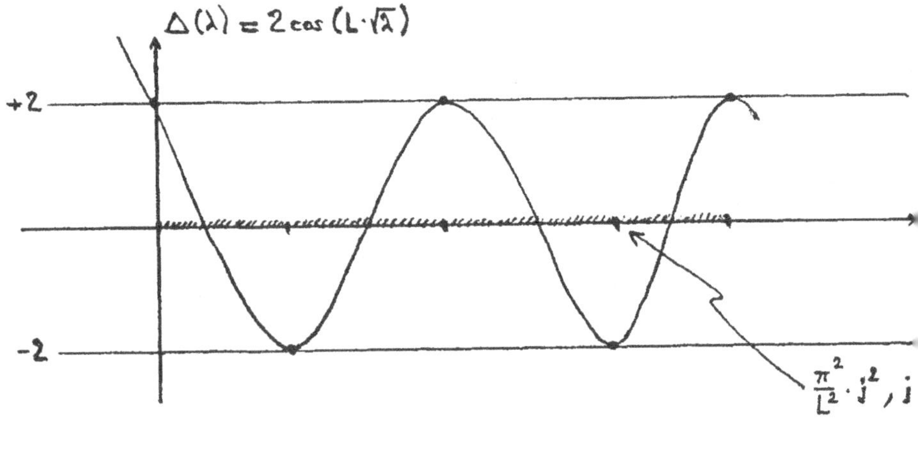

Figure 4

In the special case $q \equiv 0$ all gaps besides $(-\infty,0)$ are empty and for $\lambda \neq 0$ one has the two linearly independent solutions $g_{\pm}(\lambda;x) = e^{\pm\sqrt{-\lambda}\cdot x}$ and therefore stability if and only if $\lambda \in (0,\infty)$. In the case of general periodic potentials with period L one obtains from (2.33) – (2.36) and from Figure 3, that the graphs for the real functions $\lambda \mapsto \gamma(\lambda)$ and $\lambda \mapsto \alpha(\lambda)$ look as in Figure 5 below. The previous considerations are not enough to show the strict monotonicity of α in the interior of the bands. This property will be obtained as a corollary to more sophisticated results given in section 6, where it will be one of the main tasks to get a deeper understanding of the behaviour of the phase for general one-dimensional ergodic potentials and to explain the relation between phase and occurrence of absolutely continuous spectrum.

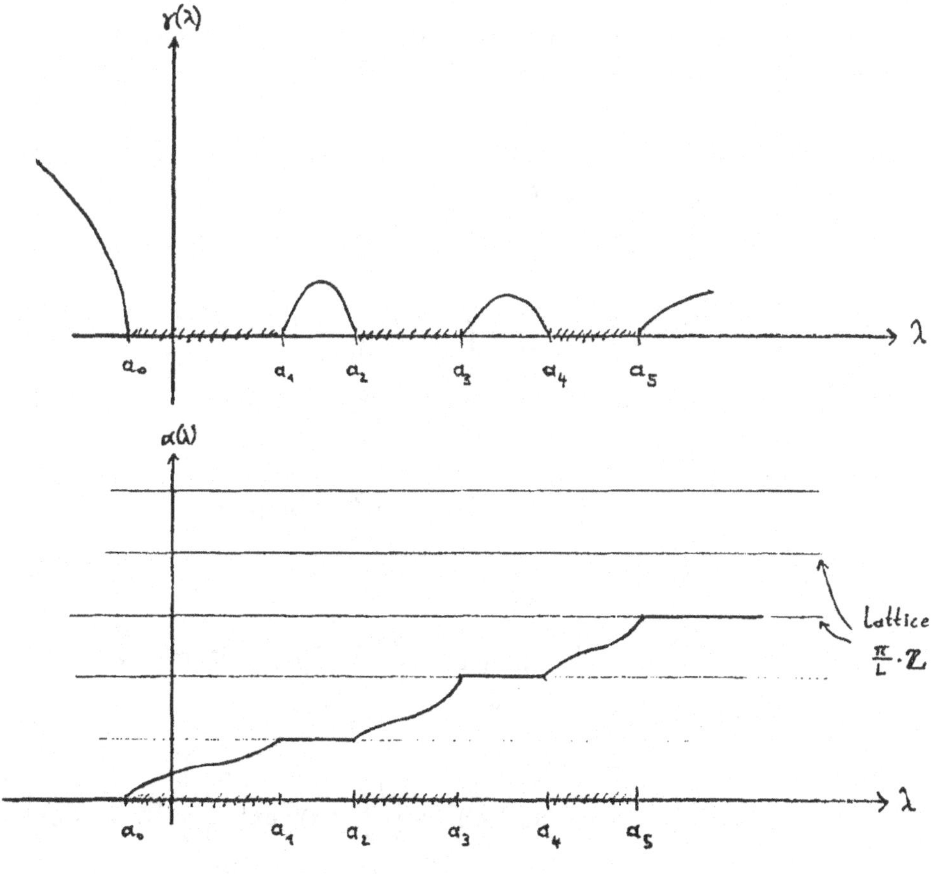

Figure 5

In particular, γ and α are not independent of each other. If $\gamma(\lambda) = 0$, λ is in a band of the spectrum and the corresponding phase varies from $\frac{\pi}{L} \cdot j$ to $\frac{\pi}{L} \cdot (j+1)$ for some $j \in \mathbf{Z}$. If γ becomes positive, the phase remains constant in $\frac{\pi}{L} \cdot \mathbf{Z}$ as λ runs through a gap in the spectrum, and so on.

We finally verify that under assumption (2.30) the operator $f \mapsto G_\lambda f$ is bounded in $L^2(\mathbf{R})$. For the proof it is sufficient to show the boundedness of the operators $K_\pm$ defined by

$$K_+f(x) = \int_x^\infty e^{-\gamma(y-x)} |f(y)| dy \ , \quad K_-f(x) = \int_{-\infty}^x e^{-\gamma(x-y)} |f(y)| dy$$

for $f \in L^2(\mathbf{R})$.

Using partial integration and the Cauchy Schwarz inequality one obtains for $L > 0$

$$\int_{-L}^{L} |K_- f(x)|^2 dx = \int_{-L}^{L} e^{-2\gamma x} \left(\int_{-\infty}^{x} e^{\gamma y} |f(y)| dy \right)^2 dx$$

$$= -\frac{1}{2\gamma} e^{-2\gamma x} \left(\int_{-\infty}^{x} e^{\gamma y} |f(y)| dy \right)^2 \Bigg|_{-L}^{+L}$$

$$+ \frac{1}{\gamma} \int_{-L}^{L} e^{-\gamma x} |f(x)| \left(\int_{-\infty}^{x} e^{\gamma y} |f(y)| dy \right) dx$$

$$\leq \frac{1}{2\gamma} \cdot (K_-f(-L))^2 + \frac{1}{\gamma} \int_{-L}^{L} |f(x)| \cdot |K_- f(x)| dx$$

$$\leq \frac{1}{2\gamma} \cdot (K_-f(-L))^2 + \frac{1}{\gamma} \left(\int_{-L}^{L} |f(x)|^2 dx \right)^{1/2} \left(\int_{-L}^{L} |K_- f(x)|^2 dx \right)^{1/2} ,$$

hence

$$\left(\int_{-L}^{L} |K_- f(x)|^2 \, dx\right)^{1/2} \leq \frac{1}{2\gamma} \cdot (K_- f(-L))^2 \cdot \left(\int_{-L}^{L} |K_- f(x)|^2 \, dx\right)^{-1/2} +$$

$$+ \frac{1}{\gamma}\left(\int_{-L}^{L} |f(x)|^2 \, dx\right)^{1/2} .$$

From the Cauchy-Schwarz inequality one easily gets $|K_- f(-L)| \to 0$ as $L \to \infty$ and therefore

$$\left(\int_{-\infty}^{\infty} |K_- f(x)|^2 \, dx\right)^{1/2} \leq \frac{1}{\gamma}\left(\int_{-\infty}^{\infty} |f(x)|^2 \, dx\right)^{1/2} .$$

$\left(\int_{-\infty}^{\infty} |K_+ f(x)|^2 \, dx\right)^{1/2}$ can be estimated similarly, and hence the boundedness of the operator $f \mapsto G_\lambda f$ is proved.

2.3. Some questions arising from the examples

Here are a few of the many questions which arise from the above examples.

(i) If one considers example 1 but with the hard core potential replaced by a continuous potential Φ, does one still have localization, even for large λ? Analogously to (2.1) one can define a Poisson model in $\mathbf{R}^d$, $d \geq 1$. What is the spectral behaviour of the corresponding H for $d > 1$? Is the higher dimensional analogue to (2.12) true for any $d \geq 1$?

(ii) How far can the concepts and results, which were introduced in section 2.2., be extended to non-periodic potentials? In particular, is it possible to generalize the characterization (2.26) of the continuous spectrum in terms of the exponent γ to the non-periodic case? How does the spectrum change, if one moves on the scale of Figure 1 from the left end, i.e. from perfectly periodic potentials, to almost periodic potentials such as

$$q(x) = A \cdot (\cos(2\pi x) + \cos(2\pi\alpha x)) \qquad (A \in \mathbf{R},\ \alpha \notin \mathbf{Q})\ ? \tag{2.37}$$

It is to be expected that the gaps become more and more dense (cf. Figure 4 and Figure 5) and that there is a tendency to a Cantor-like spectrum. Does a transition to a pure point spectrum occur, if the disorder is further increased?

(iii) The three basic quantities in the above examples are N, γ and α. Assume that these notions can be extended to the case of general one-dimensional ergodic potentials. Does a relation between N, γ and α exist?

The general heuristic picture

Guided by example 1 and 2 we try in this section to draw a general heuristic picture out the effects disorder can have on the spectral behaviour of Schrödinger operators H. We gin by describing in section 3.1. how the occurrence of multidimensional localization can be ade plausible in the case of large disorder. Here the salient feature is the absence of sonances, analogously to the absence of long range correlations in models of Statistical echanics. In section 3.2. we continue to deal with the multidimensional Poisson model and plain how the problem of the asymptotics for the integrated density of states N can be anslated into a question about the asymptotic behaviour of the paths of Brownian motion. The ost important contributions to N near the bottom of the spectrum come from large fluctuations the random potential; in terms of Brownian motion this means that large deviations from the erage behaviour of Brownian paths are essential. Finally, in section 3.3. the one-dimensional se is considered. Here, one generally expects localization for every energy value in the ectrum, if the potential contains enough disorder. The randomness entails the instability of the rresponding dynamical system, which leads to the absence of wavelike solutions and in many ses – however, exceptions do exists – to localization.

The general problem (1.2) therefore has, as we will see, at least three different faces, e which has to do with questions about thermodynamical stability, another one which is lated to the theory of large deviations for Brownian motion, and a third one which has to do ith dynamical instabilities.

1. *Multidimensional localization in the limit of large disorder*

Let $d \geq 1$ be arbitrary. In order to fix the ideas we consider the d-dimensional analogue to the Poisson model in example 1. The potential is given by

.1) $$q(x) = \sum_{i \geq 1} \Phi(x - x_i) \, ,$$

where

$$\Phi(x) = \begin{cases} +\infty \, , & |x| \leq R \\ 0 \, , & |x| > R \end{cases}$$

and the $x_i \in \mathbf{R}^d$, $i \in \mathbf{N}$, are the points of a Poisson process on $\mathbf{R}^d$ with average density 1. One can think of the x_i as centers of impurities which are assumed to be balls of a fixed radius R. Heuristically we argue that there will exist a critical value $0 < \lambda_c \leq \infty$ such that the spectrum in $[0, \lambda_c)$ is pure point and the corresponding eigenfunctions are localized, whereas the spectrum in (λ_c, ∞) is continuous and the corresponding states are extended. The reason is the following. If one recalls example 1, it seems plausible, that for small values of λ (this means that the disorder is large compared with the energy λ), the essential contributions to the spectrum come from configurations of the Poisson process, which contain large regions D_i which are free of Poisson points and whose lowest Dirichlet eigenvalue is approximately given by λ.

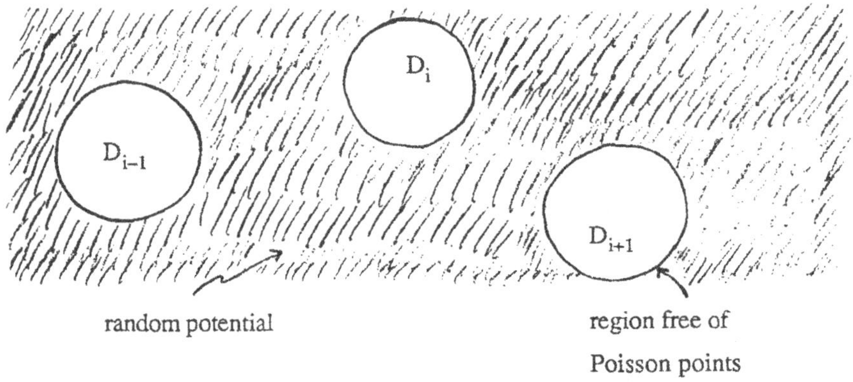

Figure 6

For the sake of simplicity let us assume that the domains D_i in Figure 6 approximately are balls. As in example 1 we denote by D(r) the open ball of radius r and by $r(\lambda)$ the radius of that ball whose lowest Dirichlet eigenvalue $\lambda_1(D(r(\lambda)))$ is λ. From the scaling relation

$$\lambda_1(D(r)) = \frac{\lambda_1(D(1))}{r^2}$$

one gets

(3.2) $$r(\lambda) \sim \lambda^{-1/2}, \quad \lambda \to 0.$$

Therefore, for a given small λ, the domains D_i in Figure 6 are balls of radius $\sim \lambda^{-1/2}$ and the corresponding eigenfunctions are approximately localized in these balls.

This heuristic consideration is convincing as long as the eigenvalue problems in the different domains D_i are essentially *separated* problems, i.e. if there are *no resonances* between the domains D_i. This should be the case if the degree of disorder is large enough respectively if λ is small. However, if the dimension d of the space is high and if λ is large, the assumption of the separation of the eigenvalue problems becomes doubtful and extended states could appear. Indeed, $\lambda_c < \infty$ is conjectured if $d \geq 3$.

This picture has some analogy with situations in Statistical Mechanics, e.g. in the Ising model, where long range correlations play a similar role as resonances do in the above model. One has roughly the following correspondences.

localization length	–	correlation length
parameter of disorder	–	temperature
mobility edge λ_c	–	critical temperature T_c
limit of large disorder	–	high temperature limit

Notice however, that this correspondence with problems in Statistical Mechanics does not extend too far; for example the free energy has a singularity at the critical temperature T_c, whereas one does not expect that the integrated density of states behaves singularly at the mobility edge.

Asymptotics for the density of states and large deviations for Brownian motion

As in the previous section we consider the Poisson model in $\mathbf{R}^d$, $d \geq 1$. The integrated density of states N is defined analogously as in the one-dimensional case. For a cube $K \subset \mathbf{R}^d$ with volume $|K|$ we denote by $H^{(K)}$ the operator, which is given by $-\Delta + q$ in the interior of K and which satisfies Dirichlet conditions on the boundary ∂K, and by $\lambda_i^{(K)}$, $i \geq 1$, the eigenvalues of $H^{(K)}$. One defines

$$N^{(K)}(\lambda) = \frac{1}{|K|} \# \{ i \in \mathbf{N} : \lambda_i^{(K)} < \lambda \}, \qquad \lambda \in \mathbf{R},$$

and

$$N(\lambda) = \lim_{K \nearrow R^d} N^{(K)}(\lambda).$$

This limit exists P – a.e. in all continuity points of N. A proof will be indicated at the end of this section. N is a deterministic distribution function and the corresponding measure will be denoted by $N(d\lambda)$.

(2.9) suggests the asymptotics

$$N(\lambda) \sim \begin{cases} \exp(-\text{const.}\ \lambda^{-d/2}) & , \lambda \to 0 , \\ \text{const.}\ \lambda^{d/2} & , \lambda \to \infty , \end{cases} \tag{3.3}$$

where ~ is used in the sense of (2.11) and (2.10) respectively. The intuitive explanation of (3.3) is the following. On the one hand it is plausible that the asymptotics of N for $\lambda \to \infty$ is the same (up to a constant factor) as in the free case $q \equiv 0$. On the other hand, according to the heuristic picture drawn in the previous section, the asymptotics of N for $\lambda \to 0$ is essentially determined by large fluctuations of the Poisson process. By an isoperimetric inequality we know that the lowest Dirichlet eigenvalue $\lambda_1(D)$ of a region D is minimal among all regions with a fixed volume, if D is a ball (an introduction to isoperimetric problems can be found for example in Chapter X in Pólya (1954)). Therefore, the essential contributions to $N(\lambda)$ come from configurations containing large balls $D(r(\lambda))$ which are free of Poisson points. Analogously to (2.12) one therefore expects

$$\begin{aligned} N(\lambda) &\sim P(D(r(\lambda)) \text{ is free of Poisson points}) \\ &= \exp(-(r(\lambda))^d) \\ &\sim \exp(-\text{const.}\ \lambda^{-d/2}) , \end{aligned} \tag{3.4}$$

where we used (3.2) in the last line.

The question of the asymptotics of $N(\lambda)$ as $\lambda \to 0$, respectively as $\lambda \to \infty$, can be translated into a question of the asymptotic behaviour of the paths $\beta(t)$ of Brownian motion as $t \to \infty$, respectively as $t \to 0$. We will explain this connection and show how the heuristic picture given above corresponds to geometric properties of Brownian paths. The heat equation will serve to link the spectrum of the Laplacian with Brownian motion. Brownian motion $\{\beta(t), t \geq 0\}$ is a random curve in $\mathbf{R}^d$ such that for $0 \leq t_0 < t_1 < \ldots < t_n$ the increments $\beta(t_i) - \beta(t_{i-1})$, $1 \leq i \leq n$, are independent of each other and Gaussian with zero mean and variance $2(t_i - t_{i-1})$. It is known that this process can be realized on the space $\mathcal{C}([0,\infty),\mathbf{R}^d)$ of continuous functions. The corresponding probability measure on $\mathcal{C}([0,\infty),\mathbf{R}^d)$ will be denoted by Q (respectively by Q_x), if the starting point of Brownian motion is $\beta(0) = 0$ (respectively $\beta(0) = x \in \mathbf{R}^d$). The transition density of Brownian motion is given by

$$p_t(x,y)dy = Q_x(\beta(t) \in dy), \qquad x,y \in \mathbf{R}^d,$$

and satisfies the heat conduction equation $\frac{\partial}{\partial t} p_t = \Delta p_t$ on the space $\mathbf{R}^d$, i.e. p_t is the kernel of the operator $e^{t \cdot \Delta}$. From this and from the spectral theorem one obtains for all $t > 0$ the following formula (proof of which we postpone to the end of this section):

(3.5) $$\int_0^\infty e^{-\lambda t} N(d\lambda) = (4\pi t)^{-d/2} \cdot P \otimes Q\,(\beta \text{ has survived until time } t \mid \beta(t) = 0).$$

Here P denotes the measure of the Poisson process on $\mathbf{R}^d$ and $\{\beta$ has survived until time $t\}$ stands for $\{\beta(s) \notin \bigcup_{i=1}^{\infty} S(x_i)$ for all $s \in [0,t]\}$, where x_i, $i \geq 1$, are the points of a realization of the Poisson process and $S(x_i)$ is the closed ball of radius R with center x_i.

Obviously the asymptotic behaviour of the right hand side of (3.5) as $t \to 0$ is given by const. $t^{-d/2}$, from which one obtains $N(\lambda) \sim$ const. $\lambda^{d/2}$ as $\lambda \to \infty$ by a Tauberian theorem. It is not so easy to find the asymptotic behaviour of the right hand side of (3.5) as $t \to \infty$. We will turn to this question in section 4.2. and in section 5. Its geometrical content is roughly the following. In order to find the rate at which the survival probability of the Brownian path tends to zero in the limit $t \to \infty$, one has to compute the weight with respect to the measure $P \otimes Q$ of those point configurations, in which a large domain around the origin is free of Poisson points, and of those paths, which hang around in this domain a very long time. The behaviour of such paths is atypical: it "deviates largely" from the behaviour of typical paths, which have a tendency to leave the domain and to dissipate in space.

We finally sketch an argument from which one obtains the convergence of the distribution functions $N^{(K)}$ and also formula (3.5). Given a cube $K \subset \mathbf{R}^d$ and a realization x_i, $i \geq 1$, of the Poisson process, we consider a Brownian motion which moves (independently of P) in the set $K \setminus \bigcup_{i=1}^{\infty} S(x_i)$ with absorption at the boundary. Its transition density, which will be denoted by $p_t^{(K)}$, formally satisfies $\frac{\partial}{\partial t} p_t^{(K)} = -H^{(K)} p_t^{(K)}$, i.e. $p_t^{(K)}$ is the kernel of $e^{-tH^{(K)}}$. For $t > 0$ fixed and for all cubes K one therefore has

$$\text{(3.6)} \quad \int e^{-\lambda t} N^{(K)}(d\lambda) = \frac{1}{|K|} \operatorname{trace}(e^{-tH^{(K)}}) = \frac{1}{|K|} \int_K p_t^{(K)}(x,x)dx =$$

$$= \frac{1}{|K|} \int_K Q_x(\beta(s) \in K \setminus \bigcup_{i=1}^{\infty} S(x_i) \quad \text{for all} \quad s \leq t \mid \beta(t) = x) \cdot (4\pi t)^{-d/2} dx .$$

In the last line of (3.6) the condition $\{\beta(s) \in K \text{ for all } s \leq t\}$ can be neglected in the limit $K \nearrow \mathbf{R}^d$, since $t > 0$ is fixed. By an application of the multidimensional ergodic theorem to the Poisson process one therefore gets for all $t > 0$ P – a.e.

$$\lim_{K \nearrow R^d} \int e^{-\lambda t} N^{(K)}(d\lambda) =$$

$$= \lim_{K \nearrow R^d} \frac{1}{|K|} \int_K Q_x(\beta(s) \notin \bigcup_i S(x_i) \quad \text{for all} \quad s \leq t \mid \beta(t) = x) \cdot (4\pi t)^{-d/2} dx$$

$$= \lim_{K \nearrow R^d} \frac{1}{|K|} \int_K Q(\beta(s) \notin \bigcup_i S(x_i - x) \quad \text{for all} \quad s \leq t \mid \beta(t) = 0) \cdot (4\pi t)^{-d/2} dx$$

$$= (4\pi t)^{-d/2} \, P \otimes Q \, (\beta \text{ has survived until time } t \mid \beta(t) = 0) .$$

From this one can conclude the existence of $\lim_{K\nearrow \mathbf{R}^d} N^{(K)}$ as well as formula (3.5) . For the proof one has to use the fact that the family $\{N^{(K)}(d\lambda),\ K\subset \mathbf{R}^d \text{ is a cube}\}$ is relatively compact in the vague topology, which can be seen as follows. One considers the free Laplacian $-\Delta$ in K with Dirichlet conditions at the boundary and denotes the corresponding eigenvalues by $\overset{\circ}{\lambda}{}_j^{(K)}$, $j \geq 1$. Since the eigenvalues of the Laplacian depend monotonically on the domain (see for example Chapter VI, §2 in Courant and Hilbert (1953)), one has $\lambda_j^{(K)} \geq \overset{\circ}{\lambda}{}_j^{(K)}$, $j \geq 1$, for all cubes K. Therefore, there exists a constant $c < \infty$ such that

$$N^{(K)}(\lambda) \leq \frac{1}{|K|}\ \#\ \{j:\ \overset{\circ}{\lambda}{}_j^{(K)} < \lambda\} \leq c \cdot \lambda^{d/2} \qquad \textit{uniformly} \text{ in K.}$$

One-dimensional localization and instability of the corresponding dynamical system

In this section we assume $d = 1$ and we give a simple heuristic explanation of localization for this special case. Let an ergodic potential be given with bounded and continuous realizations $q: \mathbf{R}^1 \to \mathbf{R}^1$. For $\lambda \in \mathbf{R}$ the equation

$$Hg_\lambda(x) = \lambda \cdot g_\lambda(x)\ , \qquad -\infty < x < +\infty\ , \tag{3.7}$$

has for a given initial condition $\begin{pmatrix} g_\lambda(0) \\ g_\lambda'(0) \end{pmatrix} = u \in \mathbf{R}^2$ a unique solution g_λ . Since $d = 1$, we can write (3.7) as a dynamical system

$$\begin{pmatrix} g_\lambda(x) \\ g_\lambda'(x) \end{pmatrix}' = \begin{pmatrix} 0 & 1 \\ q(x)-\lambda & 0 \end{pmatrix} \begin{pmatrix} g_\lambda(x) \\ g_\lambda'(x) \end{pmatrix}, \quad \begin{pmatrix} g_\lambda(0) \\ g_\lambda'(0) \end{pmatrix} = u\ , \tag{3.8}$$

whose fundamental matrix will be denoted by $Y_\lambda(x;q)$ as in section 2.2.. We want to explain the occurrence of localization with the help of the stability properties of (3.8) .

Roughly speaking, one has the following picture. As suggested by the discrete analogue to (3.8), one can imagine that the solution g_λ on $[0,\infty)$ results from the initial condition by "repeated scattering" on the potential q, where x runs from 0 to $+\infty$, and similarly on $(-\infty,0]$.

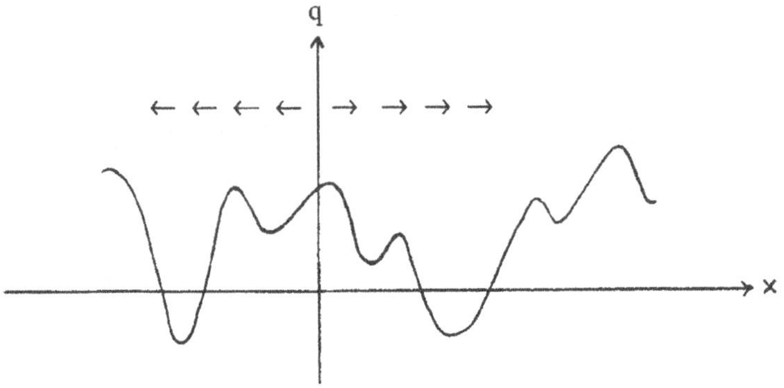

Figure 7

In the case of a periodic potential only for certain values of λ, which have to belong to one of the allowed energy bands, the solutions g_λ "fit" to the geometry of q and pass through q in the form of waves. If the minimal period of the potential tends to infinity, the gaps become more and more dense, and wavelike solutions do not exist anymore if q is sufficiently disordered.

In order to make this rough picture more precise, we begin with extending the notion of the exponent γ from the case of periodic potentials to general ergodic one-dimensional potentials. This so called (top) Ljapunov exponent is defined by

(3.9) $$\gamma(\lambda) = \lim_{x \to \pm\infty} \frac{1}{|x|} \log \| Y_\lambda(x;q) \| \quad , \qquad \lambda \in \mathbf{R} ,$$

where $\| \quad \|$ denotes the matrix norm. For fixed $\lambda \in \mathbf{R}$, this limit exists P–a.e. and is deterministic. We postpone the proof to the end of this section. Notice that $\gamma(\lambda)$ is non-negative since $\det Y_\lambda(x) = 1$ and hence $\| Y_\lambda(x) \| \geq 1$.

With the notion of Ljapunov exponent the above heuristics can be summarized as follows:

(3.10) randomness of q $\Rightarrow$ instability of (3.8)

$\Rightarrow$ positivity of γ

$\Rightarrow$ absence of wave solutions.

It is by no means a simple task to give to the first two implications of (3.10) a precise meaning. We will turn to this question in section 6. The third implication however can easily be made plausible as follows. Let $\lambda \in \mathbf{R}$ be given with $\gamma(\lambda) > 0$. This is compatible with the conservation law (2.20) only if there exist initial conditions $u_+ \in \mathbf{R}^2$ and $u_- \in \mathbf{R}^2$ such that the corresponding solutions are exponentially contracting in direction $+\infty$ and $-\infty$ respectively. Indeed, by Oseledec's multiplicative ergodic theorem (see Ruelle (1979); an informal discussion of the contents of this theorem and of its applications can be found in section 9 of Ruelle (1990)) there exist P – a.e. vectors $u_\pm = u_\pm(\lambda;q) \in \mathbf{R}^2$ such that

$$\text{11)} \quad \begin{cases} \lim_{x \to +\infty} \frac{1}{x} \log|Y_\lambda(x;q)u_+| = -\gamma(\lambda) \\ \lim_{x \to +\infty} \frac{1}{x} \log|Y_\lambda(x;q)u| = +\gamma(\lambda), \text{ if } u \in \mathbf{R}^2 \text{ is linearly independent of } u_+ \end{cases}$$

$$\text{12)} \quad \begin{cases} \lim_{x \to -\infty} \frac{1}{|x|} \log|Y_\lambda(x;q)u_-| = -\gamma(\lambda) \\ \lim_{x \to -\infty} \frac{1}{|x|} \log|Y_\lambda(x;q)u| = +\gamma(\lambda), \text{ if } u \in \mathbf{R}^2 \text{ is linearly independent of } u_- . \end{cases}$$

The solutions of (3.7) with these initial conditions $u_\pm$ will be denoted by $g_\pm(\lambda)$.

Let $\lambda \in \mathbf{R}$ be given such that $\gamma(\lambda) > 0$ and λ is not an eigenvalue. The following argument then suggests $\lambda \notin \Sigma_c$. The vectors u_+ and u_- cannot be linearly dependent since otherwise (3.11) and (3.12) would imply $g_\pm(\lambda) \in L^2(\mathbf{R}^1)$ and λ would be an eigenvalue. Hence, $g_+(\lambda)$ and $g_-(\lambda)$ are linearly independent. Let us assume for the moment, that a little bit more than (3.11) and (3.12) holds, namely

$$\text{13)} \quad |g_\pm(\lambda;x)| \leq C(\lambda) \cdot e^{\mp\gamma(\lambda)x}, \qquad -\infty < x < +\infty,$$

with a finite constant $C(\lambda)$. Then we obtain $\lambda \notin \Sigma_c$ by the same arguments as in the proof of (2.29). In general, $|g_\pm(\lambda)|$ cannot be estimated in such a strong form as in (3.13), since (3.11) and (3.12) only hold in a logarithmic sense; in section 4.3. it will be shown how one can, without assuming (3.13), slightly modify the above argument and at least rule out the possibility of absolutely continuous spectrum.

On the other hand, if λ is an eigenvalue with $\gamma(\lambda) > 0$, we have localization, since the corresponding eigenfunction is exponentially decreasing with rate $\gamma(\lambda) > 0$ because of the linear dependence of $g_+(\lambda)$ and $g_-(\lambda)$.

These heuristic arguments seem to show that positivity of the Ljapunov exponent always entails localization. Such a conclusion however is not correct in general as we will see in section 4.3. below. It turns out that the above reasoning is only sufficient to prove that positivity of the Ljapunov exponent implies absence of absolutely continuous spectrum.

To conclude this section we give the justification for (3.9). Since

$$\begin{aligned}\log \|Y_\lambda(x+y;q)\| &= \log \|Y_\lambda(y;\theta_x q) \cdot Y_\lambda(x;q)\| \\ &\leq \log\|Y_\lambda(y;\theta_x q)\| + \log \|Y_\lambda(x;q)\|,\end{aligned}$$

we can apply a subadditive ergodic theorem (there is no problem with the integrability of $\log \|Y_\lambda(x)\|$ since, for the sake of simplicity, we will assume in the following, that P is concentrated on the space of *bounded* continuous functions $q: \mathbf{R} \to [0,1]$), from which we obtain the existence $P-$a.e. of the forward limit

$$\gamma^+(\lambda) = \lim_{x \to +\infty} \frac{1}{x} \log \|Y_\lambda(x;q)\| = \inf_{x>0} \frac{1}{x} E\,[\log \|Y_\lambda(x)\|\,]$$

and of the backward limit

$$\gamma^-(\lambda) = \lim_{x \to -\infty} \frac{1}{|x|} \log \|Y_\lambda(x;q)\| = \inf_{x<0} \frac{1}{|x|} E[\log \|Y_\lambda(x)\|\,]\,.$$

Using $Y_\lambda(-x;q) = (Y_\lambda(x;\theta_{-x}q))^{-1}$ and the stationarity of the potential, we get

$$\gamma^-(\lambda) = \inf_{x>0} \frac{1}{x} E\,[\log \|\, Y_\lambda^{-1}(x)\|\,].$$

From this the equality $\gamma^-(\lambda) = \gamma^+(\lambda)$ follows since one easily verifies, using $\det Y_\lambda(x) = 1$ and Cramer's rule,

$$Y_\lambda^{-1}(x) = \begin{pmatrix} \psi_\lambda'(x) & -\psi_\lambda(x) \\ -\varphi_\lambda'(x) & \varphi_\lambda(x) \end{pmatrix}$$

$$= \begin{pmatrix} 0 & -1 \\ 1 & 0 \end{pmatrix} Y_\lambda^t(x) \begin{pmatrix} 0 & 1 \\ -1 & 0 \end{pmatrix}$$

d hence $\|Y_\lambda^{-1}(x)\| = \|Y_\lambda(x)\|$.

4. Some known results and open problems

In this section we deal with the question how far the heuristic pictures sketched in sections 3.1., 3.2. and 3.3. are mathematically founded, and some of the important open problems will be indicated. In section 4.1. a main result on multidimensional localization is stated without proof. Sections 4.2. and 4.3. are the mathematical continuations of sections 3.2. and 3.3.. Here, two Theorems are formulated, whose proofs will be explained in detail in sections 5 and 6 respectively.

4.1. Multidimensional localization in the limit of large disorder

In continuous models of Poisson type localization is not yet proved except in the trivial case of example 1, where $d = 1$ and Φ is given by a hard core potential. In dimension $d > 1$ results on localization mainly exist for discrete models on the lattice $\mathbf{Z}^d$. Prototype is the Anderson model which is defined as follows. The discrete analogue to the Laplacian is

(4.1) $$\Delta g(n) = \sum_{m \in \mathbf{Z}^d :\, |m-n|=1} g(m) \quad \text{for } g: \mathbf{Z}^d \to \mathbf{R},\ n \in \mathbf{Z}^d .$$

(To simplify the notation it is convenient to omit the diagonal term $-2d \cdot g(n)$ in the definition of the Laplacian). The Anderson Hamiltonian on $\mathbf{Z}^d$ is given by $H = -\Delta + q$, where Δ denotes the discrete Laplacian and $\{q(n),\ n \in \mathbf{Z}^d\}$ are independent identically distributed random variables. For simplicity we assume that the distribution of $q(0)$ has a bounded continuous density h. In this case, $\kappa = (\sup_{x \in \mathbf{R}} h(x))^{-1}$ can be viewed as a natural measure of the amount of disorder of the potential.

(4.2) **Result.** *Assume that the distribution of* q(0) *has a bounded continuous density* h *. Then in the Anderson model with i.i.d. potential* q(n), $n \in \mathbf{Z}^d$ $(d \geq 1)$, *the following holds. There exists* $\lambda_0 \in \mathbf{R}$ *such that the spectrum of* H *is* P–a.e. *pure point and non-empty in* $\Sigma \cap (-\infty, \lambda_0]$ *and the corresponding eigenfunctions are exponentially decreasing. If* $\kappa = (\sup_{x \in \mathbf{R}} h(x))^{-1}$ *is sufficiently large, H has P–a.e. a pure point spectrum* $\Sigma = \Sigma_{pp}$ *and all eigenfunctions are exponentially decreasing.*

Example 1 is trivial because the potential walls are infinitely high. It represents, so to speak, the case of an infinitely large disorder. If one is sufficiently near to this extreme case, localization can still be proved for certain discrete models in any dimension. This is precisely formulated in the above result. The essential step of its proof consists in making precise the analogy with problems in Statistical Mechanics indicated at the end of section 3.1.. In particular estimates of the Green's function are required analogously to estimates of the correlation function in the high temperature regime. For an explanation of the techniques of the proof we refer to Spencer (1986), Martinelli and Scoppola (1987) and to Chapter IX in Carmona and Lacroix (1990).

On a mathematical level nothing is known until now about the existence of extended states in higher dimensional space:

(4.3) **Open problem.** *If* $d > 1$ *and if the disorder is sufficiently small, do then extended states exist in the Anderson model ?*

In connection with problem (4.3) we only mention that the question of the breakdown of stability in certain two-dimensional Hamiltonian systems (see the survey article by Moser (1986)) has some similarity with the problem of the transition from extended to localized states. The analysis of a special one-dimensional model with incommensurate structures, the so called Frenkel-Kontorova model (see Aubry and Daeron (1983)), has led to new methods and results related to the breakdown of stability in two-dimensional Hamiltonian systems.

4.2. Asymptotics for the density of states and large deviations for Brownian motion

The heuristics sketched in section 3.2. is indeed correct and the following holds.

Theorem 1. *In the Poisson model* (3.1) *in* $\mathbf{R}^d$, $d \geq 1$, *with hard core potential of radius* $R > 0$ *the asymptotic behaviour of the integrated density of states* N *is given by*

$$\lim_{\lambda \to 0} \lambda^{d/2} \log N(\lambda) = -(\gamma_d)^{d/2}, \tag{4.4}$$

where γ_d *denotes the lowest Dirichlet eigenvalue of the negative Laplacian in the d-dimensional ball of unit volume .*

Remarks. (i) Theorem 1 and its proof go back (in chronological order) to Lifschitz (1965), Kac and Luttinger (1974), Donsker and Varadhan (1975 a,b,c).

(ii) For Poisson models with continuous, non-negative non-zero pair potentials Φ the asymptotics (4.4) holds also provided $\Phi(x) = o(|x|^{-d-2})$ as $|x| \to \infty$. In order to cover this more general case, only slight modifications of the proof of Theorem 1 are needed; they can be found in section 4 in Donsker and Varadhan (1975c). For simplicity of notation we restrict ourselves in the following to the case of a hard core potential.

(4.4) is equivalent to

$$\lim_{t \to \infty} t^{-d/(d+2)} \log P \otimes Q\,(\beta \text{ has survived until } t) = -c_d, \tag{4.5}$$

where the constant is independent of the potential and given by

$$c_d = \frac{d+2}{2}\left(\frac{2\gamma_d}{d}\right)^{d/(d+2)} = \frac{d+2}{2}\,|D(1)|^{2/(d+2)}\left(\frac{2\lambda_1(D(1))}{d}\right)^{d/(d+2)}.$$

This follows since a Tauberian theorem of exponential type as stated in

Fukushima (1974) implies the equivalence of (4.4) with

(4.6) $$\lim_{t\to\infty} t^{-d/(d+2)} \log \int_0^\infty e^{-\lambda t} N(d\lambda) = - c_d$$

which can be written with the help of formula (3.5) as

$$\lim_{t\to\infty} t^{-d/(d+2)} \log P \otimes Q\,(\beta \text{ has survived until } t \mid \beta(t) = 0) = - c_d\,.$$

The condition $\{\beta(t) = 0\}$ obviously does not play a role for the above asymptotics and can be omitted. The proof of Theorem 1 is therefore reduced to the proof of (4.5).

The lower bound

(4.7) $$\liminf_{t\to\infty} t^{-d/(d+2)} \log P \otimes Q\,(\beta \text{ has survived until } t) \geq - c_d$$

can easily be obtained from the following rough estimate

(4.8) $$P \otimes Q\,(\beta \text{ has survived until } t) \geq$$
$$\geq Q(\beta(s) \in D(r) \text{ for all } s \leq t) \cdot P(D(r+R) \text{ is free of Poisson points}),$$

which holds for all $r > 0$. By definition of the Poisson process the second factor on the right hand side of (4.8) equals $\exp(-|D(r+R)|)$ which is for large r of the same order as $\exp(-|D(r)|)$. In order to compute the asymptotics of the first factor as $t \to \infty$, we write

(4.9) $$Q(\beta(s) \in D(r) \text{ for all } s \leq t) = \int_{D(r)} p_t^{(r)}(0,y)dy,$$

where $p_t^{(r)}$ denotes the transition density of Brownian motion in D(r) with absorption at the boundary. We further denote by $\lambda_i(D(r))$, $i \geq 1$, the eigenvalues of the negative Laplacian in D(r) with Dirichlet boundary condition, which we assume to be ordered according to $\lambda_1(D(r)) < \lambda_2(D(r)) \leq \lambda_3(D(r)) \leq \dots$, and by $g_i^{(r)}$, $i \geq 1$, the corresponding normalized eigenfunctions. Applying the spectral theorem similarly as in (3.6) one gets

(4.10) $$p_t^{(r)}(x,y) = \sum_{i\geq 1} e^{-\lambda_i(D(r))\cdot t}\, g_i^{(r)}(x)\, g_i^{(r)}(y) \qquad (x,y \in D(r)).$$

Since $g_1^{(r)}$ is strictly positive in D(r) one obtains from (4.9) and (4.10)

(4.11) $$\lim_{t\to\infty} \frac{1}{t} \log Q(\beta(s) \in D(r) \quad \text{for all } s \leq t) = -\lambda_1(D(r)).$$

The lower bound on the right hand side of (4.8) is therefore approximately given by $\exp(-\lambda_1(D(r)) \cdot t - |D(r)|)$ as $t \to \infty$. For $t > 0$ fixed one looks for the optimal radius $r = r(t)$, for which this bound becomes as large as possible. By minimizing the function

$$r \mapsto (\lambda_1(D(r)) \cdot t + |D(r)|) = (\lambda_1(D(1)) \cdot \frac{t}{r^2} + |D(1)| \cdot r^d)$$

one obtains $r(t) = r_0 \cdot t^{1/(d+2)}$ with r_0 such that the function $r \mapsto (\lambda_1(D(r)) + |D(r)|)$ attains its minimum at r_0. The radius of the optimal ball $D(r(t))$ is therefore proportional to $t^{1/(d+2)}$ and its volume is proportional to $t^{d/(d+2)}$. This is the geometric meaning of the scaling in (4.5) with the power $d/(d+2)$, and the lower bound

(4.12) $$\begin{aligned} \liminf_{t\to\infty} \frac{1}{t^{d/(d+2)}} \log P \otimes Q \ (\beta \text{ has survived until time } t) &\geq \\ \geq -\inf_{r>0} (\lambda_1(D(r)) + |D(r)|) & \\ = -(\lambda_1(D(r_0)) + |D(r_0)|) & \\ = -c_d & \end{aligned}$$

is plausible. This heuristic argument can easily be made rigorous as follows. One scales the space with the factor $t^{-1/(d+2)}$ and uses that for any $c > 0$ the process $\{c^{-1} \cdot \beta(c^2 s), 0 \leq s\}$ is again a Brownian motion. Applying this with $c = t^{1/(d+2)}$ and denoting the closed ball of radius ρ with center $y \in \mathbf{R}^d$ by $S^\rho(y)$, one obtains

$$P \otimes Q\,(\beta(s) \notin \bigcup_i S^R(x_i) \text{ for all } s \le t)$$

$$= P \otimes Q(t^{-1/(d+2)} \cdot \beta(st^{2/(d+2)}) \notin \bigcup_i S^{Rt^{-1/(d+2)}}(x_i \cdot t^{-1/(d+2)}) \text{ for all } s \le t^{d/(d+2)})$$

$$= P \otimes Q(\beta(s) \notin \bigcup_i S^{Rt^{-1/(d+2)}}(x_i \cdot t^{-1/(d+2)}) \text{ for all } s \le t^{d/(d+2)}) .$$

Denoting by $\tau = t^{d/(d+2)}$ one therefore gets for all $r > 0$

$$\liminf_{t\to\infty} \frac{1}{t^{d/(d+2)}} \log P \otimes Q\,(\beta \text{ has survived until time t})$$

$$\ge \liminf_{\tau\to\infty} \frac{1}{\tau} \log\{Q(\beta(s) \in D(r) \text{ for all } s \le \tau) \cdot$$

$$\cdot P(x_i \cdot t^{-1/(d+2)} \notin D(r+Rt^{-1/(d+2)}) \text{ for all } i \in N)\}$$

$$= -\lambda_1(D(r)) + \liminf_{\tau\to\infty} \frac{1}{\tau} \log(\exp(-\tau \cdot |D(r+R\tau^{-1/d})|))$$

$$= -\lambda_1(Dr)) - |D(r)|$$

and the lower estimate (4.12) is verified.

It is therefore the real issue to understand, why such a rough estimate as (4.8) is already sufficient to give the exact asymptotics (4.5). In section 5 we will study this question and its relation to the heuristic picture of localization which was sketched in section 3.1..

We finally mention an unsolved question.

(4.13) **Open problem.** *Let* N *be the integrated density of states in the multidimen sional Poisson model. Is* N *a* C^∞*-function in the interior of the spectrum?*

The lack of knowledge of the regularity properties of N is related to the fact that a proof of localization in non-trivial Poisson models is still missing up to now.

4.3 The one-dimensional case

In this section we assume $d = 1$ and we turn to the question to which extent the heuristics (3.10) can be made precise and whether relation (2.26) can be extended to non-periodic potentials. The answer is formulated in Theorem 2 below. The first steps of its proof are given and further results from the one-dimenional theory are discussed. We will continue with a detailed explanation of the proof of Theorem 2 in section 6.

To understand example 1 and example 2 one had to know only the notions of pure point spectrum and of continuous spectrum which were defined with the help of eigenfunctions and approximate eigenfunctions respectively. For the following however it is necessary to refine the notion of continuous spectrum. In the one-dimensional case it is expedient to do this with the help of the spectral measure $\sigma^q(d\lambda)$, which is associated to the operator $H = H(q)$. Its definition (see for instance section 10.14 in Richtmyer (1978)) will be given in section 6.1.; in the present section one only needs to know that the spectrum is given as the support of a measure, namely that the following holds (for a proof see Richtmyer (1978), section 11.5 combined with section 10.14):

$$\begin{aligned} \Sigma(q) &= \operatorname{supp} \sigma^q \\ \Sigma_{pp}(q) &= \operatorname{supp} \sigma^q_{pp} \\ \Sigma_c(q) &= \operatorname{supp} \sigma^q_c, \end{aligned}$$

where supp denotes the support of a measure and $\sigma^q = \sigma^q_{pp} + \sigma^q_c$ is the decomposition of σ^q into its atomic and its continuous part. This definition of the spectrum allows a natural refinement of the concept of continuous spectrum. Corresponding to the Lebesgue decomposition $\sigma^q = \sigma^q_{pp} + \sigma^q_{ac} + \sigma^q_{sc}$ one defines the absolutely continuous spectrum $\Sigma_{ac}(q)$ and the singularly continuous spectrum $\Sigma_{sc}(q)$ by

$$\Sigma_{ac}(q) = \operatorname{supp} \sigma^q_{ac} \qquad \text{and} \qquad \Sigma_{sc}(q) = \operatorname{supp} \sigma^q_{sc} .$$

As suggested by the heuristics given in section 3.3., the singularly continuous spectrum should occur as an intermediate stage if the degree of order in the potential decreases from strict periodicity to strong disorder. It is exactly this possibility of a non-empty singularly continuous spectrum which is at the root of some of the difficulties of the theory to be developed below.

We further use the following notation. Let $\Omega = \mathcal{C}(\mathbf{R}, [0,1])$ be the space of continuous functions q: $\mathbf{R} \to [0,1]$, endowed with the topology of uniform convergence on compact sets, and denote by $\mathcal{F}$ the σ-algebra on Ω generated by this topology. Let P be a shift invariant ergodic probability measure on $(\Omega, \mathcal{F})$.

A random potential q is called deterministic, if the following holds for all $n \in \mathbf{N}$: q(0) is measurable with respect to the σ-algebra $\mathcal{F}_{(-\infty,-n]}$ which is generated by the variables q(x), $-\infty < x \leq -n$; or equivalently, if

$$\bigcap_{n \in \mathbf{N}} \mathcal{F}_{(-\infty,-n]} = \mathcal{F} \quad \text{up to P-negligible sets .}$$

We denote by μ the Lebesgue measure on $\mathbf{R}^1$ and for Borel sets $A \subset \mathbf{R}^1$ we write

$$\overline{A}^{\mu} = \{x \in \mathbf{R}:\ \mu(U \cap A) > 0 \text{ for any neighborhood U of x}\} .$$

Theorem 2. *Let* $(\Omega, \mathcal{F}, P)$ *be the probability space as above. Then the following holds.*

(i) q *is non-deterministic* $\Rightarrow \gamma > 0\ \ \mu -$ a.e. $\Rightarrow \Sigma_{ac} = \emptyset \quad P -$ a.e.

(ii) $\Sigma_{ac} = \overline{\{\lambda \in \mathbf{R}: \gamma(\lambda) = 0\}}^{\mu} \quad P -$ a.e. .

Remarks. (i) Theorem 2 and its proof go back (in chronological order) to Ishii (1973), Pastur (1980), Johnson and Moser (1982), Kotani (1984).

(ii) Theorem 2(i) can essentially be viewed as a precise version of (3.10). However, potentials do exist which are random from an intuitive point of view, but which nevertheless are deterministic in the sense of the definition above, for example if the pair potential Φ in (3.1) is given by a function which is analytic in a strip around the real axis. Results on the absence of absolutely continuous spectrum also in such examples can be found in Kirsch, Kotani and Simon (1985).

(iii) According to Theorem 2(ii), the Ljapunov exponent γ characterizes the absolutely continuous spectrum. From the positivity of the Ljapunov exponent one generally can conclude only $\Sigma_{ac} = \emptyset$ but not $\Sigma_c = \emptyset$ as we will see at the end of this section. It is therefore not possible to find a sufficient condition for one-dimensional localization, which would be based on the notion of Ljapunov exponent alone.

(iv) From Theorem 2(ii) the question arises for which one-dimensional potentials q the spectrum $\Sigma(q)$ is purely absolutely continuous. Owing to Theorem 2(i) such potentials q have to be deterministic. There are hints supporting the conjecture, that q necessarily has to be almost periodic, see Kotani and Krishna (1988).

(v) Only for simplicity q was assumed to be bounded and continuous. Theorem 2 holds more generally with $\Omega = L^2(\mathbf{R}; \frac{dx}{1+|x|^3})$; the proof in the general case requires only minor supplements, see Kotani (1987).

The statement of Theorem 2 can be split into two parts.

Part a. *For all Borel sets* $A \subset \mathbf{R}^1$ *one has*

$$(4.14) \qquad \gamma(\lambda) > 0 \text{ for } \mu\text{-almost all } \lambda \in A \Rightarrow \sigma^q_{ac}(A) = 0 \qquad P-\text{a.e.}.$$

Part b. *For all Borel sets* $A \subset \mathbf{R}^1$ *with* $\mu(A) > 0$ *one has*

$$(4.15) \qquad \gamma(\lambda) = 0 \quad \text{for } \mu\text{-almost all } \lambda \in A \Rightarrow \begin{cases} \dfrac{d\sigma^q_{ac}}{d\mu}(\lambda) > 0 \quad \text{for } \mu\otimes P\text{-almost all } (\lambda,q) \in A\times\Omega \\ \text{and} \\ q \text{ is deterministic}. \end{cases}$$

On the right hand side of (4.15) $\frac{d\sigma^q_{ac}}{d\mu}$ denotes the density of σ^q_{ac} with respect to the Lebesgue measure. Part b. is the difficult one; we will explain it in detail in section 6. The proof of part a. merely requires minor modifications of the heuristic argument already given in section 3.3. as we will show now.

To do so we assume

(4.16) $$\gamma(\lambda) > 0 \quad \text{for } \mu\text{-almost all } \lambda \in A,$$

where $A \subset \mathbf{R}^1$ is a given Borel set. Then (3.11) and (3.12) hold for μ-almost all $\lambda \in A$. If λ is not an eigenvalue and $\gamma(\lambda) > 0$, the argument given above (3.13) shows, that there are two linearly independent solutions $g_{\pm}(\lambda;x)$ of (3.7) which are *exponentially increasing* as $x \to \mp\infty$. In general this is not sufficient to conclude $\lambda \notin \Sigma_c$ as we could do under the assumption (3.13). However the previous argument can be modified with the help of the following growth condition (for a simple proof of which we refer for example to the proof of Lemma 4.1 in Kotani (1987)): For σ^q - almost all λ there exists a *polynomially bounded* non-trivial solution g_λ of equation (3.7).

Since this growth property holds σ^q – a.e., we cannot argue with a fixed value λ as before. This causes a (not merely technical) difficulty, because the exceptional set of potentials, for which the convergence (3.9) may fail, depends on λ and an uncountable union of null sets does not need to be again a set of measure zero. In order to overcome this difficulty one appropriately works with the product measure $\mu \otimes P$ instead with the measure P alone. Denoting by

(4.17) $$M_{\pm} = \{(\lambda,q) \in \mathbf{R} \times \Omega:\ \lim_{x \to \pm\infty} \frac{1}{|x|} \log \|Y_\lambda(x;q)\| \text{ does not exist or is different from } \gamma(\lambda)\}$$

one has

(4.18) $$\mu \otimes P\{(\lambda,q) \in A \times \Omega:\ \lambda \text{ is an eigenvalue or } (\lambda,q) \in M_{\pm}\} = 0.$$

This relation enables one to deduce properties of the measure σ^q_{ac} by using the absolute continuity

(4.19) $$P(dq)\sigma^q_{ac}(d\lambda) << P(dq)\,\mu(d\lambda)\,.$$

Using (4.16) and (4.18) one gets P – a.e.

$$(4.20) \qquad \sigma^q_{ac}(A) = \sigma^q_{ac}\{\lambda \in A : \lambda \text{ is not an eigenvalue and } \lim_{x \to \pm\infty} \frac{1}{|x|} \log \|Y_\lambda(x;q)\| = \gamma(\lambda) > 0\}.$$

Now we can argue as before and conclude that, P – a.e. for σ^q_{ac} – almost all $\lambda \in A$, there exist two linearly independent solutions growing exponentially as $x \to -\infty$ or $x \to +\infty$ respectively. Since for σ^q – almost all λ a polynomially bounded non-trivial solution exists, the right hand side of (4.20) must be zero as was to be shown.

As noticed in remark (iii), $\gamma > 0$ μ – a.e. does in general not imply $\Sigma_c = \emptyset$. The question arises, which condition in addition to the positivity of the Ljapunov exponent would be sufficient for localization. If one could work in the proof given above (cf. (4.18)) with the spectral measure σ^q itself instead merely with the Lebesgue measure μ, one indeed could deduce $\Sigma_c = \emptyset$, since the same line of reasoning as before shows that

$$(4.21) \qquad \int\int 1_{M_\pm}\, \sigma^q(d\lambda)\; P(dq) = 0$$

implies localization P – a.e..

Sufficient for (4.21) is

$$(4.22) \qquad E[\sigma(d\lambda) \mid \mathcal{F}_\pm] \ll \mu(d\lambda) \qquad P - \text{a.e.},$$

where $\mathcal{F}_+$ denotes the σ-algebra generated by $q(x)$, $x \geq 0$, and $\mathcal{F}_-$ the σ-algebra generated by $q(x)$, $x \leq 0$. Condition (4.22) can be verified for certain classes of potentials, for instance for potentials of a Markovian type, see Kotani and Simon (1987) or Chapter VIII in Carmona and Lacroix (1990). The smoothness condition (4.22) excludes the occurrence of resonances. It plays, roughly speaking, a role analogous to the condition of stability with respect to perturbations of boundary conditions in models in Statistical Mechanics.

In section 6 we will show how those aspects of Floquet theory, which are based on the notion of Floquet exponent (and which do not refer to the notion of discriminant) can be extended to general one-dimensional ergodic potentials.

Along with the proof of (4.15) we will also obtain a relation between N,γ and a generalized notion of phase α.

An extension of the Floquet representation (2.23), (2.24) to almost periodic potentials can be found in Moser and Pöschel (1984). To get results on the detailed nature of the spectrum of Hamiltonians with almost periodic potentials as (2.37) for instance, one has to use a randomization of the potential. In the continuous case there are only a few results up to now, see for example Fröhlich, Spencer and Wittwer (1990). Most other results obtained so far concern the discrete case. We conclude this section with an example which sheds light on the relation between the behaviour of the Ljapunov exponent, the occurrence of resonances and localization.

Example 3. Let a constant $A \in \mathbf{R}$ and an irrational number α be given and let ω be equidistributed in the interval [0,1]. The almost Mathieu operator on **Z** is defined by

(4.23) $$Hg(n) = -g(n-1) - g(n+1) + A \cdot \cos(2\pi(\alpha n + \omega)) \cdot g(n)$$

for $g: \mathbf{Z} \to \mathbf{R}$ and $n \in \mathbf{Z}$, and the one-sided almost Mathieu operator on **N** with boundary condition $\theta \in [0,\pi)$ is defined by

(4.24) $$\begin{aligned} &H^{\theta}g(n) = -g(n-1) - g(n+1) + A \cdot \cos(2\pi(\alpha n + \omega)) \cdot g(n), \; n \in \mathbf{N} \\ &g(0) = \cos\theta, \; g(1) = \sin\theta. \end{aligned}$$

For the almost Mathieu operator on **Z** the following holds.

(4.25) $|A| > 2 \Rightarrow \gamma(\lambda) > 0$ for all $\lambda \in \mathbf{R} \Rightarrow \Sigma_{ac} = \emptyset$ P – a.e..

(4.26) $|A| > 2$ & α is a Liouville number implies $\Sigma_{pp} = \emptyset$ P – a.e.
and therefore $\Sigma = \Sigma_{sc}$ P – a.e. because of (4.25).

An irrational number α is called a Liouville number, if there exists an approximating sequence of rationals $\frac{p_k}{q_k}$ such that $|\alpha - \frac{p_k}{q_k}| \leq k^{-q_k}$ for $k \geq 1$.

Simple proofs of (4.25) and (4.26) can be found for instance in section 10.2 of Cycon, Froese, Kirsch and Simon (1987). The last implication in (4.25) follows, since in the discrete case too positivity of the Ljapunov exponent implies absence of absolutely continuous spectrum (the proof is similar as in the continuous case). The meaning of (4.26) is the following. If $|A| > 2$ and if α can be approximated very well by rational numbers, i.e. if the potential is very near to a periodic potential, then the point spectrum of H is empty even if $\gamma(\lambda) > 0$ for all $\lambda \in \mathbf{R}$. Subtle resonances occur which lead to a singularly continuous spectrum. On the other hand, for the one-sided Mathieu operator on $\mathbf{N}$ the following holds (see Kotani (1986)):

(4.27) $|A| > 2 \Rightarrow$ for Lebesgue-almost all $\theta \in [0,\pi)$ H^{θ} has a pure point spectrum and localized eigenfunctions P – a.e..

The resonances, which can occur in the case of the Mathieu operator on $\mathbf{Z}$, are therefore excluded, if the boundary condition θ is fixed, and one has indeed localization for Lebesgue – almost all $\theta \in [0,\pi)$.

Explanation of Theorem 1 and introduction to an extended Boltzmann theory of entropy

This section is devoted to an explanation of Theorem 1. Before we begin to sketch the ›of we try to make transparent the nature of the problem. A brief description of the technical ·e of the proof follows. Then we concentrate on that part of the proof which is of general nceptual importance: In section 5.3. we give an introduction to an extended notion of ltzmann entropy and in section 5.4. we apply it to the study of large deviations for Brownian ›tion.

'. *Nature of the problem*

As explained at the end of section 4.2. (cf. (4.5) and (4.12)), for the proof of Theorem 1 we need to show

1) $$\lim_{t\to\infty} t^{-d/(d+2)} \log P \otimes Q \, (\beta \text{ has survived until time } t) \le -c_d ,$$

where

2) $$c_d = \lambda_1(D(r_0)) + |D(r_0)|$$

and

3) $$r(t) = r_0 \cdot t^{1/(d+2)}$$

is the radius of the ball $D(r(t))$ which maximizes the lower bound (4.8). In order to understand the upper bound (5.1) one has to consider two different questions.

4) *Question 1.* Why is the seemingly rough lower bound (4.8) sufficient to give the exact result ?

In order to clarify the geometric contents of this question we use the following notation.

(5.5) $W^R(t) = \{x \in \mathbf{R}^d\text{: there exists } s \in [0,t] \text{ with } |x - \beta(s)| \le R\}$

is the "Wiener sausage" of radius R swept out by the Brownian motion during the time interval [0,t]. The Brownian motion has survived until time t if and only if $W^R(t)$ is free of Poisson points. The lower bound (4.12) can be expected to be exact only if, for those paths which contribute significantly to the left hand side of (5.1), the "optimal" ball $D(r(t))$ will be nearly completely filled by the Wiener sausage $W^R(t)$ as $t \to \infty$. It has to be shown that such configurations of Brownian paths as drawn in Figure 8 for example do not contribute to the left hand side of (5.1). This is the essential step in the proof of the upper bound (5.1).

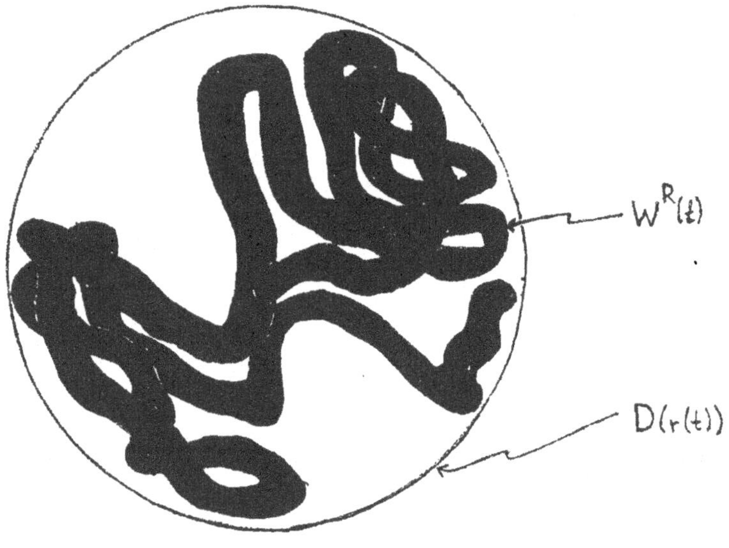

Figure 8

In the heuristic picture of section 3.1. this corresponds to the non-occurrence of resonances between different regions D_i, if λ is small.

The answer to question 1, which will be briefly discussed in section 5.2., is specific for the proof of Theorem 1. The second question is of a larger scope and its answer goes far beyond the narrow context of the original relation (5.1).

(5.6) *Question 2.* Is there a connection between the estimates (4.8), (4.12) and (5.1) on the one hand, the concept of entropy and the variational principle for the free energy in Statistical Mechanics on the other hand?

That such a relation exists is suggested by (4.8). The first factor on the right hand side of this estimate is monotonically increasing in r, the second factor is monotonically decreasing. The lower bound (4.7) is obtained by determining the optimal ratio of two opposite effects. This is a one-dimensional variational problem in which the volume $|D(r)|$ plays the role of an energy and $\lambda_1(D(r))$ plays the role of an entropy. In question 2 we ask whether this analogy can be made precise. Starting from this question we give in sections 5.3. and 5.4. an introduction to an extended Boltzmann theory of entropy and to the theory of large deviations for Brownian motion.

Relation (5.3) suggests use of a scale in which the space variable is contracted by the factor $t^{-1/(d+2)}$. In order to see what questions 1 and 2 look like on this new scale, we scale the Wiener sausage as follows:

(5.7)
$$\begin{aligned} |W^R(t)| &= \Big|\{x \in \mathbf{R}^d : \text{ there exists } s \in [0,t] \text{ with } \\ &\qquad |x-\beta(s)| \cdot t^{-1/(d+2)} \le R \cdot t^{-1/(d+2)}\}\Big| \\ &= \Big|\{x \in \mathbf{R}^d : \text{ there exists } s \in [0,t^{d/(d+2)}] \text{ with } \\ &\qquad \big|x \cdot t^{-1/(d+2)} - t^{-1/(d+2)} \cdot \beta\big(s \cdot t^{2/(d+2)}\big)\big| \le R \cdot t^{-1/(d+2)}\}\Big| . \end{aligned}$$

Using the scaling property of Brownian motion as at the end of section 4.2. one obtains from (5.7) that $|W^R(t)|$ has the same distribution as

$$\tau \cdot \left|W^{R\tau^{-1/d}}(\tau)\right| \quad \text{with } \tau = t^{d/(d+2)} .$$

Hence

(5.8)
$$\begin{aligned} &\lim_{t\to\infty} t^{-d/(d+2)} \log P \otimes Q\,(\beta \text{ has survived until time } t) = \\ &= \lim_{t\to\infty} t^{-d/(d+2)} \log E^Q \exp(-|W^R(t)|) = \\ &= \lim_{t\to\infty} \frac{1}{t} \log E^Q[\exp(-t \cdot |W^{Rt^{-1/d}}(t)|)] . \end{aligned}$$

It is easily seen that the upper bound (5.1) can hold true only if, for all $\theta > 0$,

$$\limsup_{t\to\infty} \frac{1}{t} \log Q(\,|\,W^{Rt^{-1/d}}(t)| \le |D(r_0)\,| - \theta \mid B_t) < 0 \tag{5.9}$$

where

$$B_t = \{\beta(s) \in D(r_0) \text{ for all } s \le t\}.$$

Otherwise a $\theta > 0$ would exist such that the left hand side of (5.9) would be equal to zero, i.e. with the abbreviation

$$A_t = \{|\,W^{Rt^{-1/d}}(t)\,| \le |\,D(r_0)| - \theta\}$$

one would have

$$\limsup_{t\to\infty} \frac{1}{t} \log Q\,(A_t \mid B_t) = o. \tag{5.10}$$

Using (5.10), (4.11) and (4.12) one would therefore obtain

$$\limsup_{t\to\infty} \frac{1}{t} \log E^Q \exp(-t\,|\,W^{Rt^{-1/d}}(t)\,|\,) \ge$$

$$\ge \limsup_{t\to\infty} \frac{1}{t} \log \{E^Q[1_{A_t} \cdot \exp(-t\,|\,W^{Rt^{-1/d}}(t)\,|) \mid B_t] \cdot Q(B_t)\}$$

$$\ge -\,|D(r_0)\,| + \theta + \limsup_{t\to\infty} \frac{1}{t} \log Q(A_t \mid B_t) + \lim_{t\to\infty} \frac{1}{t} \log Q(B_t)$$

$$= -\,|D(r_0)| + \theta - \lambda_1(D(r_0))$$

$$= -c_d + \theta$$

$$> -c_d$$

and (5.1) would indeed fail.

Consider the Brownian motion conditioned such that the paths do not leave the optimal ball $D(r_0)$ until time t. If the thickness $\varepsilon > 0$ of the Wiener sausage is fixed and does not shrink with increasing time, $W^{\varepsilon}(t)$ clearly tends to fill $D(r_0)$ completely as $t \to \infty$ with an error probability going to zero exponentially fast. What has to be shown is that this remains true also if $\varepsilon = \varepsilon(t)$ tends to zero as $\varepsilon(t) = R \cdot t^{-1/d}$ $(t \to \infty)$.

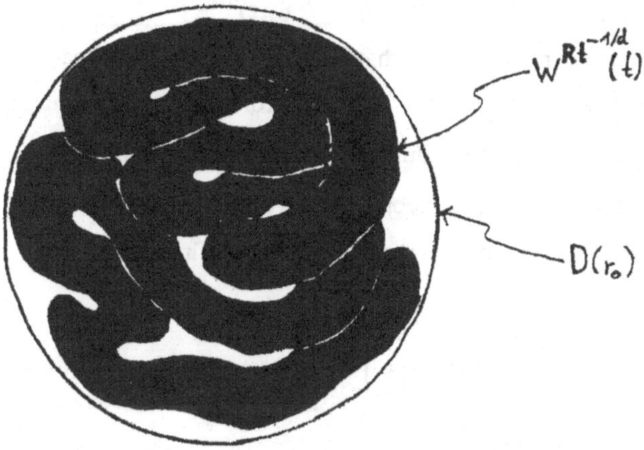

Figure 9

In this way the original problem of the non-occurrence of resonances for small λ is transformed into a geometrical question about the asymptotic behaviour of the paths of Brownian motion as $t \to \infty$.

In order to answer this question, one has – and here the connection with question (5.6) appears – to "count" those paths $\{\beta(s), 0 \leq s \leq t\}$, whose support, thickened by a ball of radius $\varepsilon(t)$, has a volume of a given value. The study of the asymptotic behaviour of the average volume of the Wiener sausage was initiated by Spitzer (1964). However, since one must now consider in particular paths whose behaviour deviates substantially from the average behaviour, one is beyond the realm of the law of large numbers. What we intend to do now is to show the connection between this problem and the notion of entropy. We begin with the remark that it is not essential for the following discussion, whether we deal in (5.9) with the Brownian motion $\{\beta(s), 0 \leq s \leq t\}$ under the condition B_t or whether we consider a Brownian motion on a compact state space, say a torus. The only essential point is that we consider a Brownian motion which does not dissipate in the whole space $\mathbf{R}^d$, but remains in a bounded set. In order to define what Brownian motion on a torus precisely means, we use the following notation. Let $K \subset \mathbf{R}^d$ be a cube with side length L and denote by $T^{(K)} = \mathbf{R}^d / L \cdot \mathbf{Z}^d$ the corresponding flat torus, i.e. we identify two points in $\mathbf{R}^d$ with each other if their coordinates differ only by integer multiples of L. The Brownian motion $\beta^{(K)}$ on $T^{(K)}$ is defined as the projection of the Brownian motion β on $\mathbf{R}^d$ to $T^{(K)}$. If it starts in 0 we denote the corresponding probability measure on $\mathcal{C}([0,\infty), T^{(K)})$ by $Q^{(K)}$.

Next we make precise what we mean by "counting" paths and define the analogue to Boltzmann's "occupation numbers" for paths of Brownian motion. We denote by $\mathcal{M}^{(K)}$ the space of probability measures on $T^{(K)}$ and interpret measures $\rho \in \mathcal{M}^{(K)}$ as macrostates. On the microscopic level we consider the occupation measure $L_t^{(K)}$ $(t > 0)$ of Brownian motion $\{\beta(s), 0 \leq s \leq t\}$ which is defined by

$$L_t^{(K)}(B) = \frac{1}{t}\int_0^t 1_B(\beta^{(K)}(s))ds \qquad \text{for Borel sets } B \subset T^{(K)}. \tag{5.11}$$

The question is "how many" paths of Brownian motion are compatible with a given macrostate $\rho \in \mathcal{M}^{(K)}$ if t is large, i.e. one asks for

$$Q^{(K)}\,(L_t^{(K)} \approx \rho) \underset{t\to\infty}{\sim} \;? \tag{5.12}$$

In order to formulate the answer we denote by $\mu^{(K)}$ the Lebesgue measure on $T^{(K)}$ and define the functional $I^{(K)}: \mathcal{M}^{(K)} \to [0,\infty]$ by

$$I^{(K)}(\rho) = \begin{cases} \int |\nabla\psi(x)|^2 \mu^{(K)}(dx), & \text{if } \rho = \psi^2 \cdot \mu^{(K)} \\ \\ +\infty, & \text{if } \rho \text{ is not absolutely continuous with respect to } \mu^{(K)}. \end{cases} \tag{5.13}$$

In sections 5.3. and 5.4. we will show that

$$Q^{(K)}(L_t^{(K)} \approx \rho) \sim e^{-I^{(K)}(\rho)\cdot t}, \quad t \to \infty, \tag{5.14}$$

and we will explain the following correspondence between (5.11), (5.13), (5.14) and analogous notions in Statistical Mechanics

$L_t^{(K)}$	– occupation numbers
$I^{(K)}$	– negative entropy
$Q^{(K)}(L_t^{(K)} \approx \rho) \sim e^{-I^{(K)}(\rho)\cdot t}$	– entropy = log (probability).

We finally argue on a heuristic level why it is the Dirichlet integral (5.13) which plays the role of a negative entropy in (5.14). For continuous functions $V: T^{(K)} \to \mathbf{R}$ we consider the operator $H = -\Delta + V$ on the flat torus $T^{(K)}$ and we denote by $\lambda(V)$ its lowest eigenvalue. The heat equation with cooling term V

$$\begin{cases} \frac{\partial}{\partial t} p_t^V(x,y) = -\ H p_t^V(x,y) = \Delta p_t^V(x,y) \ - V(x)\cdot p_t^V(x,y) \\ \lim\limits_{t\to 0} p_t^V(x,y) = \delta(x-y) \end{cases} \tag{5.15}$$

gives a connection between the spectrum of H and the Brownian motion $\beta^{(K)}$, since by the Feynman-Kac formula the solution to (5.15) can be represented as a Wiener integral

$$p_t^V(x,y) = E_x^{(K)}\left[\exp\left(-\int_0^t V(\beta^{(K)}(s))ds\right);\ \delta(\beta^{(K)}(t)-y)\right], \tag{5.16}$$

where $E_x^{(K)}$ denotes the expectation with respect to the Brownian motion on $T^{(K)}$ starting in $x \in T^{(K)}$. Expanding the left hand side of (5.16) analogously to (4.10) one obtains for the lowest eigenvalue $\lambda(V)$ of H the expression

$$\lambda(V) = -\lim_{t\to\infty} \frac{1}{t} \log E_x^{(K)}\left[\exp\left(-\int_0^t V(\beta^{(K)}(s))ds\right)\right]. \tag{5.17}$$

On the other hand, as is known from classical analysis, $\lambda(V)$ can be characterized by a variational principle, the so called Rayleigh - Ritz principle

$$\lambda(V) = \inf \{ \int |\nabla\psi|^2 d\mu^{(K)} + \int |\psi|^2 V d\mu^{(K)} : \psi \in L^2(T^{(K)}), \int |\psi|^2 d\mu^{(K)} = 1 \}. \tag{5.18}$$

Putting (5.17) and (5.18) together one obtains

$$\lim_{t\to\infty} \frac{1}{t} \log E_x^{(K)} \left[\exp \left(- \int_0^t V(\beta^{(K)}(s))ds \right) \right] = \tag{5.19}$$

$$= -\inf \left\{ \int |\nabla\psi|^2 d\mu^{(K)} + \int |\psi|^2 V d\mu^{(K)} : \psi \in L^2(T^{(K)}), \int |\psi|^2 d\mu^{(K)} = 1 \right\}$$

and the question arises whether one can understand the variational formula (5.19) directly by means of properties of the Wiener measure alone without relying on the Rayleigh-Ritz principle. It is this problem that Kac refers to on p. 196 in his seminal work Kac (1951) with the remark "It is curious that the classical variational expression (5.18) does not seem to imply or be implied directly by (5.17)." In sections 5.3. and 5.4. we try to explain the answer found by Donsker and Varadhan (1975 a).

If one assumes – in the following one of the main tasks will consist in making precise this assumptions and in justifying it –, that for large t the probability $Q^{(K)}(L_t^{(K)} \approx \rho)$ can be written in the form (5.14) with an unknown functional $I^{(K)}$ which has still to be determined, then (5.19) suggests that this functional has to be the Dirichlet integral. This can heuristically be seen as follows. We denote by $\mathcal{C}(T^{(K)})$ the space of continuous functions on $T^{(K)}$ and use the abbreviation

$$\langle \rho, V \rangle = \int V d\rho \quad \text{for } \rho \in \mathcal{M}^{(K)} \text{ and } V \in \mathcal{C}(T^{(K)}). \tag{5.20}$$

As will become clear from (5.22) below, the topology on $\mathcal{M}^{(K)}$, which is appropriate for our purposes, is the topology of weak convergence. A sequence $\rho_n \in \mathcal{M}^{(K)}$ is said to converge weakly to $\rho \in \mathcal{M}^{(K)}$, if and only if

21) $$\langle \rho_n, V \rangle \to \langle \rho, V \rangle \qquad \text{as } n \to \infty$$

for all $V \in C(T^{(K)})$. The compactness of $T^{(K)}$ entails the weak compactness of $\mathcal{M}^{(K)}$. If one covers $\mathcal{M}^{(K)}$ by finitely many small open sets U_i, $1 \le i \le m$, one gets with the help of (5.14) approximately

22) $$E_0^{(K)}\left[\exp\left(-\int_0^t V(\beta^{(K)}(s))ds\right)\right]$$

$$= E_0^{(K)}\left[\exp\left(-t\left\langle L_t^{(K)}, V\right\rangle\right)\right]$$

$$\approx \sum_{i=1}^{m} Q^{(K)}\left(L_t^{(K)} \in U_i\right) \cdot \exp\left(-t \cdot \inf_{\rho \in U_i} \cdot \langle \rho, V \rangle\right)$$

$$\approx \sum_{i=1}^{m} \exp\left(-t \cdot I^{(K)}(\rho_i)\right) \cdot \exp\left(-t \cdot \inf_{\rho \in U_i} \langle \rho, V \rangle\right)$$

where $\rho_i \in U_i (1 \le i \le m)$. In the last line of (5.22) the term with the largest exponent dominates asymptotically as $t \to \infty$. If the covering is fine enough, (5.22) therefore suggests the asymptotics

23) $$\lim_{t \to \infty} \frac{1}{t} \log E_0^{(K)}\left[\exp\left(-\int_0^t V(\beta^{(K)}(s))ds\right)\right]$$

$$= -\inf_{\rho \in \mathcal{M}^{(K)}} \left\{I^{(K)}(\rho) + \langle \rho, V \rangle\right\}.$$

Comparing (5.23) with (5.19) one can indeed identify the functional $I^{(K)}$ with the Dirichlet integral (5.13).

5.2. The technical core of the proof

This section may be skipped by readers interested mainly in the conceptual part of the proof of (5.1). Here we formulate the main lemma, on which the answer to question (5.4) is based. To do so we introduce the notion of a mollifier k_ε of radius $\varepsilon > 0$ and the convolution $k_\varepsilon * \rho$ of such a mollifier with measures $\rho \in \mathcal{M}^{(K)}$. Let a C^∞-function $k: \mathbf{R}^d \to [0,\infty)$ be given which satisfies $\{x \in \mathbf{R}^d : k(x) > 0\} = \{x \in \mathbf{R}^d: |x| < 1\}$, $\int k(x)dx = 1$ and $k(-x) = k(x)$. For $\varepsilon > 0$ we define $k_\varepsilon(x) = \varepsilon^{-d} k(x/\varepsilon)$ and

$$k_\varepsilon * \rho(x) = \sum_{g \in L \cdot Z^d} \int k_\varepsilon(x + g - y)\, \rho(dy) \qquad \text{for } \rho \in \mathcal{M}^{(K)}.$$

The L^1-norm of a function f on the torus $T^{(K)}$ is denoted by $\| f \|_1 = \int |f(x)|\, \mu^{(K)}(dx)$.

<u>**Main Lemma.**</u> *Let* $\varepsilon: [0,\infty) \to (0,\infty)$ *be given such that*

(5.24) $$\liminf_{t \to \infty} \varepsilon(t) \cdot t^{1/d} > 0 .$$

Then for all $\theta > 0$

(5.25) $$\limsup_{\delta \to 0} \limsup_{t \to \infty} \frac{1}{t} \log Q^{(K)} \left(\| k_\delta * k_{\varepsilon(t)} * L_t^{(K)} - k_{\varepsilon(t)} * L_t^{(K)} \|_1 \geq \theta \right) = -\infty .$$

For the proof, which is rather technical, we refer to Donsker and Varadhan (1975b) and Varadhan (1984); a different proof for the upper bound (5.1) can be found in Sznitman (1990). Here we content ourselves to show how the main lemma enables one to answer question (5.4), and to indicate the first step on the way which leads from (5.25) to the proof of (5.1).

Let K be fixed and $\varepsilon(t) = R \cdot t^{-1/d}$. The analogue to (5.9) for the Brownian motion $\beta^{(K)}$ on $T^{(K)}$ is

(5.26) $$\limsup_{t \to \infty} \frac{1}{t} \log Q^{(K)} \left(\left| T^K \setminus W^{\varepsilon(t)}(t) \right| \geq \theta \right) < 0$$

for all $\theta > 0$. We want to show how (5.26) is implied by (5.25). Denoting by $f^{(K)} = \frac{1}{|K|} \cdot 1_{T^{(K)}}$ we use

27)
$$| T^{(K)} \setminus W^{\varepsilon(t)}(t) | = \int \left(1_{T^{(K)}}(x) - 1_{W^{\varepsilon(t)}(t)}(x)\right) \mu^{(K)}(dx)$$

$$\leq |K| \cdot \int | f^{(K)}(x) - k_{\varepsilon(t)} * L_t^{(K)}(x)| \; \mu^{(K)}(dx)$$

$$= |K| \cdot \| f^{(K)} - k_{\varepsilon(t)} * L_t^{(K)} \|_1$$

and furthermore for all $\delta > 0$

28)
$$\{ \| f^{(K)} - k_{\varepsilon(t)} * L_t^{(K)} \|_1 \geq \theta \} \subset$$

$$\subset \{ \| f^{(K)} - k_\delta * k_{\varepsilon(t)} * L_t^{(K)} \|_1 \geq \frac{\theta}{2} \} \cup \{ \| k_\delta * k_{\varepsilon(t)} * L_t^{(K)} - k_{\varepsilon(t)} * L_t^{(K)} \|_1 \geq \frac{\theta}{2} \}.$$

It is intuitively plausible and not difficult to prove that for fixed $\delta > 0$

29)
$$\limsup_{t \to \infty} \frac{1}{t} \log Q^{(K)} \left(\| f^{(K)} - k_\delta * k_{\varepsilon(t)} * L_t^{(K)} \|_1 \geq \frac{\theta}{2} \right) < 0$$

independently of the velocity with which $\varepsilon(t)$ tends to zero. In order to handle the second term on the right hand side of (5.28), condition (5.24) becomes essential. From (5.27), (5.28), (5.29) and (5.25) one obtains the desired estimate (5.26).

We finally indicate the set-up for the proof of

0)
$$\limsup_{t \to \infty} \frac{1}{t} \log E^Q \exp(-t |W^{\varepsilon(t)}(t)|) \leq - \inf_{r > 0} \{ | D(r) | + \lambda_1 (D(r)) \}.$$

Since for all cubes K

1)
$$E^Q \exp(-t | W^{\varepsilon(t)}(t) |) \leq E^{Q^{(K)}} \exp(-t | W^{\varepsilon(t)}(t) |),$$

the essential part of the proof of (5.30) consists in finding the appropriate upper bound for $\limsup_{t\to\infty} \frac{1}{t} \log E^{Q^{(K)}} \exp(-t\,|W^{\varepsilon(t)}(t)|)$ with fixed K.

For probability densities f on $T^{(K)}$ let us define a functional Φ by

$$\Phi(f) = |\{\, x \in T^{(K)}: f(x) > 0 \,\}|\,. \tag{5.32}$$

If U_i, $1 \le i \le m$, is a covering of $\mathcal{M}^{(K)}$ by finitely many sets, one can estimate

$$\begin{aligned} (5.33)\qquad & E^{Q^{(K)}} \exp(-t\,|W^{\varepsilon(t)}(t)|) \\ & \le E^{Q^{(K)}} \exp\left(-t\,\Phi\left(k_{\varepsilon(t)} * L_t^{(K)}\right)\right) \\ & \le \sum_{i=1}^{m} Q^{(K)}\left(k_{\varepsilon(t)} * L_t^{(K)} \in U_i\right) \cdot \exp\left(-t \cdot \inf_{\rho\in U_i} \Phi(\rho)\right). \end{aligned}$$

One now wants to continue similarly as in (5.22). For this purpose the sets U_i, $1 \le i \le m$, have to be "small" with respect to a topology which is fine enough so that Φ becomes lower semicontinuous. This is the case for the topology of L^1-convergence but it is no longer true for the analogously to (5.21) defined weak topology on $L^1(T^{(K)})$. In the general theory to be developed in section 5.4. the asymptotics (5.14) will be made precise by giving lower bounds of $Q^{(K)}(L_t^{(K)} \in G)$ for weakly open sets $G \subset \mathcal{M}^{(K)}$ and upper bounds of $Q^{(K)}(L_t^{(K)} \in A)$ for weakly closed sets $A \subset \mathcal{M}^{(K)}$. In (5.33) we need analogous upper bounds for $Q^{(K)}(k_{\varepsilon(t)} * L_t^{(K)} \in A)$, but for sets $A \subset L^1(T^{(K)})$ which are closed in the L^1-topology. Here the main lemma comes in. As is shown in Donsker and Varadhan (1975b) and in Varadhan (1984), the lemma allows us to convert upper bounds involving weakly closed sets in corresponding upper bounds involving L^1-closed sets. The upshot is that one can indeed continue the estimate (5.33) in the heuristically indicated way and one finally obtains the upper bound

$$\limsup_{t\to\infty} \frac{1}{t} \log E^{Q^{(K)}} \exp\left(-t \,|W^{a(t)}(t)|\,\right) \le$$

$$\le \; -\inf_{\rho \in \mathcal{M}^{(K)}} \left\{ I^{(K)}(\rho) + |\operatorname{supp} \rho| \right\}$$

fixed K. Passing to the limit $K \nearrow \mathbf{R}^d$ one obtains from the last estimate by standard uments (see Donsker and Varadhan (1975c)) the upper bound

$$-\inf\{\lambda_1(G) + |G| : \; G \subset \mathbf{R}^d \text{ is open and has a smooth boundary}\},$$

ch, by means of an isoperimetric inequality for the lowest Dirichlet eigenvalue, reduces to one-dimensional variational formula on the right hand side of (5.30).

Introduction to an extended Boltzmann theory of entropy

We begin with the original formula of Boltzmann which is obtained by counting and by an application of Stirling's formula. Let n particles be contained in a fixed volume which we think to be partitionned into r cells. Let the cells be numbered by i, $1 \le i \le r$. We assume that the particles are distributed independently of each other and that the probability of a particle being in the cell with number i is $\frac{1}{r}$ $(1 \le i \le r)$. We will interpret probability measures $\rho = (\rho_1, \rho_2, \ldots \rho_r)$ on the set $\{1, 2, \ldots, r\}$ as macro-states with the meaning of ρ_i as the occupation density of the cell with number i $(1 \le i \le r)$. Denoting by W the probability distribution of the n-particle configuration and setting the Boltzmann constant equal to 1, the entropy S of a macrostate ρ is computed by

4) $$S(\rho) = \lim_{n\to\infty} \frac{1}{n} \log W\{\text{ configurations which are compatible with } \rho\}$$

$$= \lim_{n\to\infty} \frac{1}{n} \log W\{\text{ the cell with number i contains approximately } n_i \approx n \cdot \rho_i \text{ particles}, \; 1 \le i \le r\}$$

$$= \lim_{n\to\infty} \frac{1}{n} \log \left(\frac{n!}{n_1! \ldots n_r!} \left(\frac{1}{r}\right)^n \right)$$

$$= \lim_{n\to\infty} \frac{1}{n} \log \frac{n^n}{(n\rho_1)^{n\rho_1} \ldots (n\rho_r)^{n\rho_r}} \quad - \log r$$

$$= -\sum_{i=1}^{r} \rho_i \log \rho_i \; - \log r \,.$$

How can this simple consideration be extended to more general distributions W in a way which finally enables one to "count occupation numbers" of the paths of Brownian motion? The method to be used will consist, as we will see, in replacing direct counting by inequalities, similarly as with generalizations of the weak law of large numbers from binomially distributed variables to more general distributions. However such inequalities have now to be more refined since the entropy has to be defined in particular for states far from equilibrium where one is out of the realm of the law of large numbers.

Before we come to Brownian motion itself in the next section, we consider the following context which is slightly more general than in the example given above. Let $\mathcal{X}$ be a compact metric space, $\Omega = \mathcal{X}^{\mathbf{Z}}$ and let X_k: $\Omega \to \mathcal{X}$ be the coordinate map $X_k(\omega) = \omega_k$ $(k \in \mathbf{Z})$, where $\omega = (\omega_k)_{k \in \mathbf{Z}} \in \Omega$. The empirical distribution of the variables X_k, $1 \le k \le n$, is denoted by

(5.35) $$L_n = \frac{1}{n} \sum_{k=1}^{n} \delta_{X_k} ,$$

i.e. $L_n(A) = \frac{1}{n}\ \# \{k \le n \colon X_k \in A\}$ for Borel sets $A \subset \mathcal{X}$. We further denote by $\mathcal{M} = \mathcal{M}(\mathcal{X})$ the space of probability measures on $\mathcal{X}$ and by

(5.36) $$Q_\rho = \bigotimes_{\mathbf{Z}} \rho$$

the infinite product measure with identical factors $\rho \in \mathcal{M}$. Let a measure $\pi \in \mathcal{M}$ be given. In the following we consider Q_π as a fixed reference measure on Ω. For $\rho \in \mathcal{M}$ we ask for the probability

(5.37) $$Q_\pi (L_n \approx \rho) \sim \ ?$$

asymptotically as $n \to \infty$.

The example in the beginning corresponds to the special case with $\mathcal{X} = \{1,2, \dots , r\}$, X_k = number of the cell, in which the particle with number k is placed, $L_n(i)$ = fraction of particles in the cell numbered by i, and with the reference measure π given by $\pi_i \equiv \frac{1}{r}$ $(1 \le i \le r)$. According to (5.34) the answer to (5.37) in this special case is $Q_\pi(L_n \approx \rho) \sim e^{S(\rho) \cdot n}$, $n \to \infty$. In the general case the answer to (5.37) is

8) $$Q_\pi(L_n \approx \rho) \sim e^{-h(\rho|\pi)\cdot n}\ , \quad n \to \infty\ ,$$

where the relative entropy (= information gain) of a measure $\rho \in \mathcal{M}$ with respect to the measure $\pi \in \mathcal{M}$ is defined by

9) $$h(\rho \,|\, \pi) \;=\; \begin{cases} \displaystyle\int_{X} \log \frac{d\rho}{d\pi}\, d\rho, & \text{if } \rho << \pi \\ +\infty & \text{, otherwise .} \end{cases}$$

Here $\rho << \pi$ means that ρ is absolutely continuous with respect to π and $\frac{d\rho}{d\pi}$ denotes the density (Radon-Nikodym derivative) of ρ with respect to π. In the above example one has $h(\rho|\pi) = -S(\rho)$, and generally h has the meaning of a negative entropy.

In order to give the symbol $\approx$ in (5.38) an exact meaning we have to endow $\mathcal{M}(X)$ with a topology. The appropriate one turns out to be the topology of weak convergence which is defined as at the end of section 5.1. (cf. (5.21)). We use the same notation as there, replacing the base space $T^{(K)}$ by X; in particular we write $\langle \rho, V \rangle = \int V d\rho$ for $\rho \in \mathcal{M}(X)$ and $V \in \mathcal{C}(X)$. The precise formulation of (5.38) is the following.

Proposition 1. *Let* $G \subset \mathcal{M}(X)$ *be open in the weak topology.*
Then

0) $$\liminf_{n\to\infty} \frac{1}{n} \log Q_\pi\big(L_n \in G\big) \;\geq\; -\inf_{\rho\in G} h(\rho|\pi)\ .$$

Let $A \subset \mathcal{M}(X)$ *be closed in the weak topology. Then*

1) $$\limsup_{n\to\infty} \frac{1}{n} \log Q_\pi\big(L_n \in A\big) \;\leq\; -\inf_{\rho\in A}\ h(\rho|\pi)\ .$$

We give the main steps of the proof of Proposition 1 in such a way that they can be extended immediately to much more general situations.

Step 1. Lower bound.

Let an open set $G \subset \mathcal{M}(\mathcal{X})$ and $\rho \in G$ with $\rho \ll \pi$ be given. We have to show

(5.42) $$\liminf_{n\to\infty} \frac{1}{n} \log Q_\pi\big(L_n \in G\big) \geq -h(\rho\,|\,\pi) .$$

By the strong law of large numbers, L_n converges weakly to ρ Q_ρ – a.e. and therefore

(5.43) $$Q_\rho(L_n \in G) \to 1 \qquad \text{as } n \to \infty ,$$

since G is open and $\rho \in G$. However the reference measure in (5.42) is Q_π and not Q_ρ. Regarding an application of the law of large numbers, Q_π is the "wrong" reference measure. It is therefore natural to transform Q_π into the appropriate measure Q_ρ and to try to measure the amount of this transformation (by the way, this is exactly what the Greek word entropy means, namely "content of transformation"). We denote by $\mathcal{F}_n$ the σ-algebra generated by $X_1, X_2, \ldots, X_n$ and by $\left.\frac{dQ_\rho}{dQ_\pi}\right|_{\mathcal{F}_n}$ the Radon-Nikodym derivative of the measure Q_ρ restricted to $\mathcal{F}_n$ with respect to the measure Q_π restricted to $\mathcal{F}_n$. Then we obtain

(5.44) $$Q_\pi(L_n \in G) \geq \int_{\{L_n \in G\}} 1_{\left\{\left.\frac{dQ_\rho}{dQ_\pi}\right|_{\mathcal{F}_n} > 0\right\}} \, dQ_\pi$$

$$= \int_{\{L_n \in G\}} \left(\left.\frac{dQ_\rho}{dQ_\pi}\right|_{\mathcal{F}_n}\right)^{-1} dQ_\rho$$

$$= \int_{\{L_n \in G\}} \exp\left(-\log \left.\frac{dQ_\rho}{dQ_\pi}\right|_{\mathcal{F}_n}\right) dQ_\rho .$$

Now we can apply the law of large numbers with respect to the "right" measure Q_ρ and we can use the Q_ρ–a.e. convergence

(5.45)
$$\frac{1}{n}\log\frac{dQ_\rho}{dQ_\pi}\bigg|_{\mathcal{F}_n} = \frac{1}{n}\sum_{k=1}^{n}\log\frac{d\rho}{d\pi}(X_k)$$

$$\to E^\rho\left[\log\frac{d\rho}{d\pi}\right] = h(\rho|\pi)\ , \quad n\to\infty .$$

From (5.44) we get for $\varepsilon > 0$

$$Q_\pi(L_n\in G) \geq \int_{\{L_n\in G\}} 1\left\{\frac{1}{n}\log\frac{dQ_\rho}{dQ_\pi}\bigg|_{\mathcal{F}_n} \leq h(\rho|\pi)-\varepsilon\right\}\cdot\exp\left(-\frac{dQ_\rho}{dQ_\pi}\bigg|_{\mathcal{F}_n}\right) dQ_\rho$$

$$\geq Q_\rho\left(\{L_n\in G\}\cap\left\{\frac{1}{n}\sum_{k=1}^{n}\log\frac{d\rho}{d\pi}(X_k)\leq h(\rho|\pi)-\varepsilon\right\}\right)\cdot$$

$$\cdot\ \exp(-n(h(\rho|\pi)-\varepsilon))$$

and using (5.43) and (5.45) we conclude the desired lower bound (5.42).

<u>Step 2.</u> Upper bound.

Let $V\in\mathcal{C}(\mathcal{X})$ be given. Similarly to the Chebyshev inequality one can estimate

(5.46)
$$Q_\pi(L_n\in A) = \int_{\{L_n\in A\}} \exp(n\langle L_n, V\rangle - n\langle L_n, V\rangle)\, dQ_\pi$$

$$\leq \int \exp(n\cdot\sup_{\rho\in A}\langle\rho, V\rangle - n\langle L_n, V\rangle)\, dQ_\pi .$$

Using the notation

(5.47)
$$\lambda(V) = -\lim_{n\to\infty}\frac{1}{n}\log E^{Q_\pi}\exp(-n\langle L_n, V\rangle) = -\log E^\pi[e^{-V}]$$

one gets from (5.46)

$$\limsup_{n\to\infty} \frac{1}{n} \log Q_\pi \left(L_n \in A\right) \leq \sup_{\rho\in A} \langle \rho, V \rangle - \lambda(V) \tag{5.48}$$

for all $V \in \mathcal{C}(X)$, hence

$$\limsup_{n\to\infty} \frac{1}{n} \log Q_\pi \left(L_n \in A\right) \leq \inf_{V\in \mathcal{C}(X),\, \lambda(V)=0} \ \sup_{\rho\in A} \langle \rho, V \rangle \tag{5.49}$$

for all closed $A \subset \mathcal{M}(X)$. If $\mathcal{M}(X)$ would be a finite set, one could obviously interchange inf and sup on the right hand side of (5.49). Since $\mathcal{M}(X)$ is compact, a standard covering argument, which will be given at the end of this section, shows that such an interchange is indeed justified in the case at hand. Hence

$$\limsup_{n\to\infty} \frac{1}{n} \log Q_\pi \left(L_n \in A\right) \leq - \inf_{\rho\in A} \hat{h}(\rho|\pi) \tag{5.50}$$

with

$$\hat{h}(\rho|\pi) = \inf_{V\in \mathcal{C}(X),\, \lambda(V)=0} \langle \rho, V \rangle = - \inf_{V\in \mathcal{C}(X)} \left\{ \langle \rho, V \rangle - \lambda(V) \right\} \tag{5.51}$$

This yields the upper bound (5.41) with $\hat{h}$ instead of h.

<u>Step 3.</u> Proof of the equality

$$\hat{h}(\rho|\pi) = h(\rho|\pi) \ . \tag{5.52}$$

We first note that the function $\rho \to h(\rho|\pi)$ is convex and non-negative and that it is zero if and only if $\rho = \pi$. This can easily be seen for example by writing $h(\rho|\pi) = E^\pi[\psi\left(\frac{d\rho}{d\pi}\right)]$ with the non-negative, strictly convex function $\psi(t) = t \log t - t + 1$.

We keep the reference measure π fixed. If we think of $V \in \mathcal{C}(\mathcal{X})$ as a potential, the equilibrium distribution with respect to this potential is, as usual in Statistical Mechanics, given by

$$\pi^V = \frac{e^{-V}}{E^{\pi} e^{-V}} \cdot \pi . \tag{5.53}$$

Here we have set the temperature equal to 1. For a given V, it is the equilibrium measure π^V, which is the natural reference measure, and not π. This suggests computing the entropy $h(\rho|\pi)$ with the help of the following transformation formula

$$\begin{aligned} h(\rho|\pi) &= \int \log \frac{d\rho}{d\pi}\, d\rho = \int \log \frac{d\rho}{d\pi^V}\, d\rho + \int \log \frac{d\pi^V}{d\pi}\, d\rho \\ &= h(\rho|\pi^V) - \langle \rho, V \rangle + \lambda(V) . \end{aligned} \tag{5.54}$$

Since $h(\rho|\pi^V) \geq 0$, (5.54) implies $h(\rho|\pi) \geq \hat{h}(\rho|\pi)$. To show the reverse inequality, one defines for a given ρ a potential V by $V = -\log \frac{d\rho}{d\pi}$ (if this expression does not exist or fails to be in $\mathcal{C}(\mathcal{X})$, one has to approximate $\log \frac{d\rho}{d\pi}$ appropriately by $\mathcal{C}(\mathcal{X})$-functions; we omit this merely technical point). Then $\rho = \pi^V$ and $\lambda(V) = 0$ and from (5.54) one obtains the desired inequality $h(\rho|\pi) = -\langle \rho, V \rangle \leq \hat{h}(\rho|\pi)$.

As simple as this proof of (5.52) is, the actual reason behind the coincidence of the lower bound proved in step 1 and the upper bound proved in step 2 remains so far a little mysterious. The key for an understanding of (5.52) comes from the convexity of the function $\rho \mapsto h(\rho|\pi)$. By (5.54) one easily obtains the Gibbsian variational principle

$$\lambda(V) = \inf_{\rho \in \mathcal{M}} \{\langle \rho, V \rangle + h(\rho|\pi)\} , \quad V \in \mathcal{C}(\mathcal{X}) , \tag{5.55}$$

and given $V \in \mathcal{C}(\mathcal{X})$, the free energy $\rho \mapsto (\langle \rho, V \rangle + h(\rho|\pi))$ takes its minimum exactly for the Gibbs measure $\rho = \pi^V$. Equation (5.55) can be written in the form

$$-\lambda(-V) = \sup_{\rho \in \mathcal{M}} \{\langle \rho, V \rangle - h(\rho|\pi)\} . \tag{5.56}$$

This means that $V \mapsto -\lambda(-V)$ is just the Legendre transform h^* of the convex function h (see Figure 10 below). The actual reason for the equality (5.52) is therefore the involution property $h^{**} = h$ which says

$$h(\rho|\pi) = h^{**}(\rho|\pi) = \sup_{V\in \mathcal{C}(\mathcal{X})} \{\langle \rho, V\rangle - h^*(\rho|\pi)\} \tag{5.57}$$

$$= \sup_{V\in \mathcal{C}(\mathcal{X})} \{\langle \rho, V\rangle + \lambda(-V)\}$$

$$= -\inf_{V\in \mathcal{C}(\mathcal{X})} \{\langle \rho, V\rangle - \lambda(V)\}$$

$$= \hat{h}(\rho|\pi)\,.$$

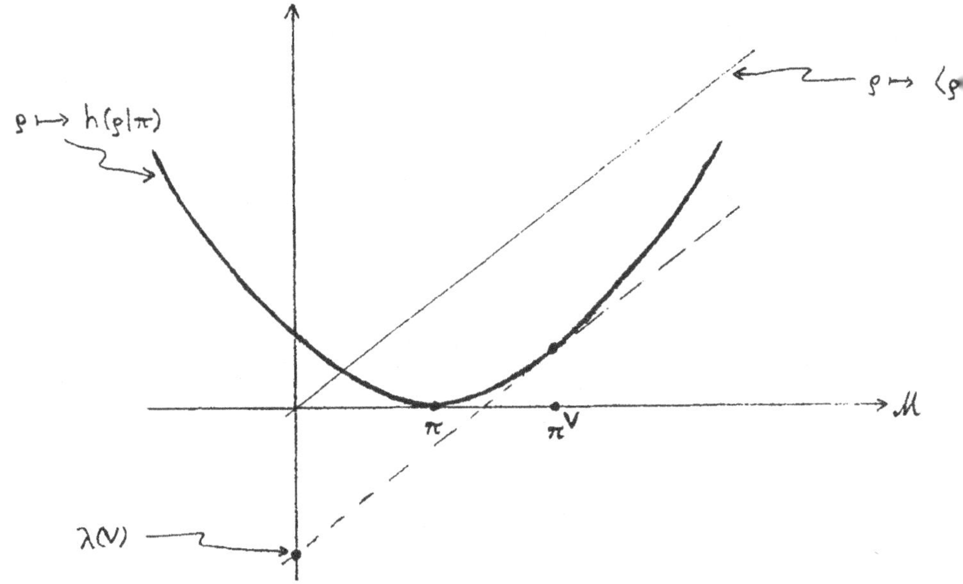

Figure 10

To conclude this section we prove the interchangeability of inf and sup on the right hand side of inequality (5.49). According to (5.49) one has for all compact $A \subset \mathcal{M}(\mathcal{X})$

$$\limsup_{n\to\infty} \frac{1}{n} \log Q_\pi(L_n \in A) \le \inf_{V\in \mathcal{V}} \sup_{\rho\in A} \langle \rho, V\rangle$$

with $\mathcal{V} = \{V \in \mathcal{C}(X): \lambda(V) = 0\}$. Given a compact set $A \subset \mathcal{M}(X)$ and $\varepsilon > 0$ one has to show

$$\limsup_{n\to\infty} \frac{1}{n} \log Q_\pi(L_n \in A) \le L + \varepsilon ,$$

where $L = \sup_{\rho \in A} \inf_{V \in \mathcal{V}} \langle \rho, V \rangle$.

For any covering of A by finitely many compact neighborhoods N_j, $1 \le j \le k$, one obtains from (5.49)

$$\begin{aligned}
\limsup_{n\to\infty} \frac{1}{n} \log Q_\pi(L_n \in A) &\le \limsup_{n\to\infty} \frac{1}{n} \log \left\{ \sum_{j=1}^{k} Q_\pi(L_n \in N_j) \right\} \\
&\le \max_{1 \le j \le k} \limsup_{n\to\infty} \frac{1}{n} \log Q_\pi(L_n \in N_j) \\
&\le \max_{1 \le j \le k} \inf_{V \in \mathcal{V}} \sup_{\rho \in N_j} \langle \rho, V \rangle .
\end{aligned}$$

Hence it is enough to find a covering of A by finitely many compact neighborhoods N_j, $1 \le j \le k$, such that

$$\inf_{V \in \mathcal{V}} \sup_{\rho \in N_j} \langle \rho, V \rangle \le L + \varepsilon , \quad 1 \le j \le k .$$

According to the definition of L, for any $\mu \in A$ there exists a potential $V_\mu \in \mathcal{V}$ with $\langle \mu, V_\mu \rangle < L + \varepsilon$ for all $\rho \in N_\mu$. Finitely many of these sets N_μ are enough to cover the compact set A. Such a covering has the desired property and the interchange of inf and sup on the right hand side of (5.49) is therefore justified.

5.4. Large deviations for Brownian motion

Here we come back to the original question (5.12) and we indicate how the previous considerations naturally lead to its answer. The reference measure is now the measure $Q^{(K)}$ of Brownian motion on the torus $T^{(K)}$. In the following the cube K is fixed and for simplicity the upper index K will be dropped. We consider the torus $T = T^{(K)}$ and the space $\mathcal{M} = \mathcal{M}(T)$ of probability measures on T, endowed with the topology of weak convergence. The empirical distribution L_t of Brownian motion on T and the Dirichlet integral I are defined as in (5.11) and (5.13). With these notations the following holds.

Proposition 2. *Let* $G \subset \mathcal{M}$ *be open in the weak topology. Then*

$$\liminf_{t\to\infty} \frac{1}{t} \log Q\left(L_t \in G\right) \geq -\inf_{\rho \in G} I(\rho) . \tag{5.58}$$

Let $A \subset \mathcal{M}$ *be closed in the weak topology. Then*

$$\limsup_{t\to\infty} \frac{1}{t} \log Q\left(L_t \in A\right) \leq -\inf_{\rho \in A} I(\rho) . \tag{5.59}$$

The proof follows exactly the scheme developed in the previous section. We indicate the main steps and refer to Donsker and Varadhan (1975a) for technical details.

Step 1. Lower bound.

Let $\rho \in G$ be given. For simplicity we assume that ρ has a strictly positive and smooth density f with respect to the Lebesgue measure μ on T (otherwise one has to approximate ρ appropriately by smooth measures and to modify the following proof correspondingly). Obviously Brownian motion on T has the Lebesgue measure as invariant measure and $L_t \to \mu$ Q – a.e. . Given $\rho = f{\cdot}\mu$, we look for a diffusion process Q^f on T with a drift still to be determined. We want to choose it in such a way that $\rho = f{\cdot}\mu$ becomes the invariant measure of the diffusion process and we therefore have

0) $$L_t \to f\cdot\mu \quad Q^f - \text{a.e.}\,.$$

If the drift is the negative gradient of a potential $\Phi\colon T \to \mathbf{R}$, which means that the generator of the corresponding diffusion process on the flat torus T is given by

1) $$L = -\left(\nabla\Phi\right)\cdot\nabla + \Delta\,,$$

then the corresponding invariant measure is just the Gibbs measure with respect to the potential Φ (cf. (5.53))

2) $$\frac{e^{-\Phi}}{\left\langle \mu, e^{-\Phi}\right\rangle}\cdot\mu\,,$$

since one easily computes $L^*(e^{-\Phi}) = 0$ for the adjoint operator $L^* = \nabla\left(\left(\nabla\Phi\right)\cdot\right) + \Delta$. We apply this with $\Phi = -\log f$ and obtain that the diffusion process with generator

3) $$L^f = (\nabla(\log f))\cdot\nabla + \Delta$$

has $f\cdot\mu$ as invariant measure. Let Q^f be the probability measure on $\mathcal{C}([0,\infty),T)$ which corresponds to the diffusion process with generator (5.63) and which starts in the point 0. By an ergodic theorem we then have (5.60) with this measure Q^f. Because of $\rho = f\cdot\mu \in G$ one obtains from (5.60)

4) $$Q^f(L_t \in G) \to 1 \text{ as } t \to \infty\,.$$

Denoting by $\mathcal{F}_t$ the σ-algebra generated by $\{\beta(s), 0 \le s \le t\}$ and proceeding in the same way as in (5.44) before, one gets

5) $$Q(L_t \in G) \ge \int_{\{L_t\in G\}} \exp\left(-\log \frac{dQ^f}{dQ}\bigg|_{\mathcal{F}_t}\right) dQ^f\,.$$

With some stochastic calculus (Cameron - Martin formula) one can compute

(5.66) $$\lim_{t\to\infty} \frac{1}{t} \log \frac{dQ^f}{dQ}\Big|_{\mathcal{F}_t} = I(f\cdot\mu) \quad Q^f - \text{a.e.}\ .$$

Using (5.64) and (5.66) one obtains from (5.65) the desired lower bound

$$\liminf_{t\to\infty} \frac{1}{t} \log Q(L_t \in G) \geq - I(f\cdot\mu)\ .$$

<u>Step 2.</u> Upper bound.

Analogously to (5.46) one gets for all $V \in \mathcal{C}(T)$

(5.67) $$Q(L_t \in A) = \int_{\{L_t \in A\}} \exp\left(t\langle L_t, V\rangle - t\ \langle L_t, V\rangle\right) dQ$$

$$\leq \exp(t \cdot \sup_{\rho\in A} \langle \rho, V\rangle\ ) \cdot E^Q\Big[\exp - \int_0^t \left(V(\beta(s)\right) ds)\Big]$$

and with

(5.68) $$\lambda(V) = -\lim_{t\to\infty} \frac{1}{t} \log E^Q\Big[\exp\Big(-\int_0^t V(\beta(s))\, ds\Big)\Big]$$

one obtains for all $V \in \mathcal{C}(T)$

(5.69) $$\limsup_{t\to\infty} \frac{1}{t} \log Q(L_t \in A) \leq - \sup_{\rho\in A} \langle \rho, V\rangle - \lambda(V)\ .$$

By the same argument as in step 2 in the previous section we get from (5.69)

(5.70) $$\limsup_{t\to\infty} \frac{1}{t} \log Q(L_t \in A) \leq - \inf_{\rho\in A} \hat{I}(\rho)$$

with

(5.71) $$\hat{I}(\rho) = - \inf_{V\in\mathcal{C}(T)} \{\langle\rho, V\rangle - \lambda(V)\} = - \inf_{V\in\mathcal{C}(T), \lambda(V)=0} \langle \rho, V\rangle\ .$$

To obtain the last equality in (5.71) we replace V by $(V - \lambda(V))$. This transformation to "right" potentials V satisfying $\lambda(V) = 0$ resembles the transformation of Q to a "right" measure in step 1 above.

<u>Step 3.</u> Proof of the equality

72)
$$\hat{I}(\rho) = I(\rho) \ .$$

In order to make formula (5.71) more transparent we have to recall the analytical meaning of $\lambda(V)$, thereby coming back to our original theme of spectral theory. By the Feynman-Kac formula we know (recall (5.17)) that $\lambda(V)$ is the lowest eigenvalue of the operator $H = -\Delta + V$. For $V \in \mathcal{C}(T)$ with $\lambda(V) = 0$ the ground state u of H satisfies the eigenvalue equation

3)
$$(-\Delta + V)\, u \;=\; \lambda(V) \cdot u = 0.$$

According to the theorem of Perron - Frobenius, u is strictly positive and hence V can be expressed in terms of u by

74)
$$V = \frac{\Delta u}{u} \ .$$

With (5.74) and $\rho = f \cdot \mu$ one can write (5.71) in the form

5)
$$\hat{I}(\rho) = -\inf_{u \in \mathcal{C}_+^2} \left\langle \rho \,,\, \frac{\Delta u}{u} \right\rangle \;=\; -\inf_{u \in \mathcal{C}_+^2} \int_T \frac{\Delta u(x)}{u(x)}\, f(x)\, \mu(dx) \,,$$

where $\mathcal{C}_+^2 = \{\, u \in \mathcal{C}(T)$: u is twice continuously differentiable and strictly positive $\}$. In particular for $u = \sqrt{f}$ we obtain from (5.75)

$$\hat{I}(\rho) \geq -\int_T \sqrt{f}\, \Delta(\sqrt{f})\, d\mu \;=\; \int_T \left| \nabla(\sqrt{f}) \right|^2 d\mu = I(\rho) \ .$$

To see the reverse inequality we write $u \in \mathcal{C}_+^2$ in the form $u = e^h$ and by using partial integration we compute

$$-\int \frac{\Delta u}{u} f\, d\mu = -\left\{\int |\nabla h|^2 f\, d\mu + \int (\Delta h) f\, d\mu\right\}$$

$$= -\int \left(\sqrt{f}\,|\nabla h|\right)^2 d\mu + 2\int \left(\sqrt{f}\,\nabla h\right)^2 \cdot \nabla(\sqrt{f})\, d\mu$$

$$\leq \int |\nabla(\sqrt{f})|^2\, d\mu = I(\rho)\,.$$

This proves (5.72). We finally notice that relation (5.75) also shows the convexity of the functional $\hat{I}$ on $\mathcal{M}$.

As in the previous section, the geometrical reason behind (5.72) is the involution property of the Legendre transformation. The Gibbsian variational principle (5.55) is now replaced by the Rayleigh-Ritz principle (5.18) for the lowest eigenvalue $\lambda(V)$ of the operator $H = -\Delta + V$, which can be written in the form

$$\text{(5.76)} \qquad \lambda(V) = \inf_{\psi \in L^2(T),\ \int |\psi|^2 d\mu = 1} \left\{\int |\nabla\psi|^2\, d\mu + \int |\psi|^2 V\, d\mu\right\}$$

$$= \inf_{\rho \in \mathcal{M}} \{\, I(\rho) + \langle \rho, V\rangle \,\}\,.$$

Hence $-\lambda(-\,.\,)$ is the Legendre transform I^* of the convex functional I and equality (5.72) is nothing else than the involution property $I^{**} = I$.

From the Legendre transform we can also see how the proof of Proposition 2 and the original heuristic considerations, which were based on the Rayleigh-Ritz principle (cf. (5.23)), are related to each other. On the other hand the theory developed so far can be seen as far reaching generalization of this principle as follows. The argument sketched in (5.22) can easily be made precise with the help of Proposition 2. Namely, for any covering of $\mathcal{M}$ with closed neighborhoods A_i, $1 \leq i \leq m$, one obtains

7) $$-\lambda(V) = \lim_{t\to\infty} \frac{1}{t} \log E^Q \left[\exp\left(-\int_0^t V(\beta(s))\,ds\right)\right]$$

$$\leq \limsup_{t\to\infty} \frac{1}{t} \log \sum_{i=1}^{m} \int_{A_i} Q(L_t \in d\rho)\, \exp\left(-t \cdot \lim_{\rho\in A_i} \langle \rho, V \rangle\right)$$

$$\leq \max_{1\leq i\leq m} \left\{ -\inf_{\rho\in A_i} I(\rho) \;-\; \inf_{\rho\in A_i} \langle \rho, V \rangle \right\} .$$

Because of (5.71) I is lower semicontinuous, and since the neighborhoods A_i can be chosen arbitrarily small, (5.77) implies

8) $$-\lambda(V) \leq -\inf_{\rho\in \mathcal{M}} \left\{ I(\rho) + \langle \rho, V \rangle \right\} .$$

On the other hand one can estimate for any $\varepsilon > 0$ and for sufficiently small open sets $G \subset \mathcal{M}$

9) $$-\lambda(V) \geq \liminf_{t\to\infty} \frac{1}{t} \log \int_G Q(L_t \in d\rho)\, \exp\left(-t \cdot \left(\inf_{\rho\in G} \langle \rho, V \rangle + \varepsilon\right)\right)$$

$$\geq -\inf_{\rho\in G} I(\rho) \;-\; \inf_{\rho\in G} \langle \rho, V \rangle - \varepsilon .$$

If we choose $\rho_0 \in \mathcal{M}$ so that

$$I(\rho_0) + \langle \rho_0, V \rangle \leq \inf_{\rho\in\mathcal{M}} \{ I(\rho) + \langle \rho, V \rangle \} + \varepsilon$$

and if G is chosen as a sufficiently small neighborhood of ρ_0, one obtains from (5.79)

0) $$-\lambda(V) \geq -\inf_{\rho\in\mathcal{M}} \left\{ I(\rho) + \langle \rho, V \rangle \right\}$$

and because of (5.78), (5.76) is deduced. In this way the Rayleigh-Ritz principle is implied by the theory of large deviations for Brownian motion, and the original remark by Kac mentioned after formula (5.19) is clarified. Looking back at the proof of a precise version of the heuristic formula (5.14), a remark of Schrödinger about the structure of statistical thermodynamics comes to mind. On p. 36/37 in Schrödinger (1946) he writes:" One of the fascinating features of statistical thermodynamics is that quantities and functions, introduced primarily as mathematical devices, almost invariably acquire a fundamental physical meaning."

Explanation of Theorem 2 and introduction to an extended Floquet-Weyl theory

In this section the proof of the remaining part (4.15) of Theorem 2 will be given. We suppose d = 1 and consider one-dimensional ergodic potentials. For the sake of simplicity we assume that the potential belongs to the space Ω of continuous functions $q: \mathbf{R} \to [0,1]$. In sections 6.1 and 6.2 we recall some notions from the spectral theory of one-dimensional Schrödinger operators and try to explain the basic ideas of the proof of (4.15). Section 6.3. contains an introduction to an extended Floquet-Weyl theory. Finally, in section 6.4. we sketch some remaining parts of the proof of Theorem 2.

Nature of the problem and some notions from spectral theory

According to Figure 5, in the case of periodic potentials the phase α is strictly increasing in the interior of the intervals where the Ljapunov exponent γ vanishes. We now ask what we can learn from Figure 5 for the general case of non-ergodic potentials and how this picture is related to the statement of Theorem 2 that vanishing of γ entails positivity of the absolutely continuous part of the spectral measure.

Before we try to answer this question, we give the definition of the spectral measure and of the quantities N, α and γ for the case of general one-dimensional ergodic potentials belonging to the class described above. In order to define the spectral measure corresponding to the operator $H = -\frac{d^2}{dx^2} + q$, we use the following notation. The resolution of the identity of H is denoted by $\mathcal{E}(d\lambda)$ and its integral kernel by $\mathcal{E}(d\lambda;x,y)$; this means $H = \int \lambda\, \mathcal{E}(d\lambda)$ and $\int \mathcal{E}(d\lambda;x,y) = \delta(x-y)$. For a bounded interval $K \subset \mathbf{R}^1$, the operator $H^{(K)}$ is defined by $H^{(K)} = -\frac{d^2}{dx^2} + q$ on the interior of K with Dirichlet boundary conditions. The resolution of the identity of $H^{(K)}$ is denoted by $\mathcal{E}^{(K)}(d\lambda)$ and its kernel by $\mathcal{E}^{(K)}(d\lambda;x,y)$. $H^{(K)}$ has a discrete spectrum. We denote the corresponding eigenvalues and eigenfunctions by $\lambda_i^{(K)}$ and $g_i^{(K)}$ respectively ($i \geq 1$). Provided that the $g_i^{(K)}$ are normalized to $\| g_i^{(K)} \| = (\int | g_i^{(K)}(x)|^2 dx)^{1/2} = 1$, the kernel $\mathcal{E}^{(K)}(d\lambda;x,y)$ can be expressed by means of the eigenvalues and eigenfunctions as follows:

$$\mathcal{E}^{(K)}(d\lambda;x,y) = \sum_{i\geq 1} \delta_{\lambda_i^{(K)}}\, g_i^{(K)}(x)\, g_i^{(K)}(y) \qquad (x,y \in \mathbf{R}).$$

Writing the functions $g_i^{(K)}$ as linear combinations of the solutions φ_λ and ψ_λ introduced in section 2.2., below formula (2.17), one obtains from (6.1) in the limit $K \uparrow \mathbf{R}$ the following representation of $\mathcal{E}(d\lambda;x,y)$ (cf. section 10.14 in Richtmyer (1978))

(6.2)
$$\mathcal{E}(d\lambda;x,y) = e^q_{11}(d\lambda)\,\varphi_\lambda(x)\,\varphi_\lambda(y) + e^q_{12}(d\lambda)\varphi_\lambda(x)\,\psi_\lambda(y) + e^q_{21}(d\lambda)\psi_\lambda(x)\varphi_\lambda(y) + e^q_{22}(d\lambda)\,\psi_\lambda(x)\,\psi_\lambda(y)\,,$$

where

$$\begin{pmatrix} e^q_{11}(d\lambda) & e^q_{12}(d\lambda) \\ e^q_{21}(d\lambda) & e^q_{22}(d\lambda) \end{pmatrix}$$

is a 2×2 - matrix valued measure. The spectral measure of H is defined as the trace of this matrix:

(6.3)
$$\sigma^q(d\lambda) = e^q_{11}(d\lambda) + e^q_{22}(d\lambda).$$

As noticed already in section 4.3., the spectral measure and the spectrum of H are related by

(6.4)
$$\begin{aligned} \Sigma_{ac}(q) &= \operatorname{supp}\ \sigma^q_{ac} \\ \Sigma_{sc}(q) &= \operatorname{supp}\ \sigma^q_{sc} \\ \Sigma_{pp}(q) &= \operatorname{supp}\ \sigma^q_{pp}\ . \end{aligned}$$

It turns out that the measure $e^q_{11}(d\lambda)$ already contains the information about the spectral measure which we need for the following considerations. We give a special name to this measure and write

(6.5)
$$e^q_{11}(d\lambda) = \rho(d\lambda;q).$$

From now on we refer to $\rho(d\lambda;q)$ as the spectral measure of H. This notational inconsistency should not cause trouble since in the following we are working merely with the measure ρ and do not rely on the full spectral matrix.

The integrated density of states N is defined as in section 3.2. by

$$N = \lim_{K \uparrow \mathbf{R}} N^{(K)} ,$$

where

$$N^{(K)}(\lambda) = \frac{1}{|K|} \quad \#\{\, i \in \mathbf{N}:\ \lambda_i^{(K)} < \lambda \} \qquad , \lambda \in \mathbf{R} ,$$

and $\lambda_i^{(K)}$, $i \geq 1$, are the eigenvalues of $H^{(K)}$. The limit (6.6) exists P – a.e. in the continuity points of N (it turns out that N is always continuous in dimension $d = 1$, see step 5 in the proof of Proposition 5 in section 6.3. below). This can be proved similarly as in (3.6) since q is bounded from below. According to the Feynman - Kac formula (5.16), the integral kernel of e^{-tH} can be represented as

$$\int e^{-\lambda t}\, \mathcal{E}(d\lambda; x,y) = E_x^Q \,[\exp(-\int_0^t q(\beta(s))ds);\ \delta(\beta(t) - y)] \qquad (x,y \in \mathbf{R}),$$

where E_x^Q denotes the expectation with respect to Brownian motion on $\mathbf{R}$ starting in x. Proceeding analogously to (3.6) one obtains, this time with the help of the Feynman - Kac formula, the existence of the limit (6.6) and the formula

$$\int e^{-\lambda t}\, N(d\lambda) = E^P\, E_0^Q \,[\exp(-\int_0^t q(\beta(s))\, ds)\ ;\ \delta(\beta(t))\,] , \quad t > 0.$$

From this one can easily deduce a relation between the spectral measure and the integrated density of states as follows. The operator e^{-tH} $(t > 0)$ has the kernel $\int e^{-\lambda t} \mathcal{E}(d\lambda;x,y)$ and therefore (6.7) and (6.8) imply

$$\int e^{-\lambda t} N(d\lambda) = E \int e^{-\lambda t}\, \mathcal{E}(d\lambda;0,0) \ , \qquad t > 0.$$

By (6.2) and (6.5) one has $\mathcal{E}(d\lambda;0,0) = \rho(d\lambda;q)$. Hence it follows from (6.9)

$$N(d\lambda) = E[\rho(d\lambda;q)]$$

and in this sense the measure $\rho(d\lambda;q)$ plays the role of an analogue to $N(d\lambda)$ defined pointwise for individual q.

To introduce the notion of the phase $\alpha(\lambda)$, $\lambda \in \mathbf{R}$, we recall example 2.2.. In the case of general non-periodic potentials, a period $L < \infty$ is no more available. (2.21) and (2.22) suggest to generalize the notion of the Floquet exponent w by taking, so to say, the limit of an infinite period and to define

$$w(\lambda) = - \lim_{L \to \infty} \frac{1}{L} \log g_\lambda(L) \tag{6.11}$$

for appropriate solutions g_λ of equation (1.3), $\lambda \in \mathbf{R}$. Given (6.11) it is natural to define the phase α, analogously to (2.25), by

$$\alpha(\lambda) = - \operatorname{Im} \lim_{L \to \infty} \frac{1}{L} \int_0^L \frac{d}{dx} \log g_\lambda(x)\,dx = \tag{6.12}$$

$$= - \lim_{L \to \infty} \frac{1}{L} \int_0^L \operatorname{Im} \frac{g_\lambda'(x)}{g_\lambda(x)}\,dx \,, \qquad \lambda \in \mathbf{R} \,,$$

where g_λ' denotes the derivative of g_λ with respect to x. In order to find a condition on g_λ which guarantees the meaningfulness of the right hand side of (6.12), we remember that the Wronskian (3.19) is a conserved quantity. Let λ be real and g_λ a solution of (1.3). Then $\overline{g}_\lambda$ is again a solution and hence $[g_\lambda,\overline{g}_\lambda](x)$ is independent of x. Under the assumption $[g_\lambda,\overline{g}_\lambda](0) \neq 0$ the integrand on the right hand side of (6.12) is therefore well defined since then g_λ has no zeroes. For definiteness we assume that g_λ is a (necessarily complex) solution satisfying

$$\operatorname{Im} [g_\lambda,\overline{g}_\lambda](0) > 0 \,. \tag{6.13}$$

Notice that $[g_\lambda,\overline{g}_\lambda]$ is purely imaginary. Because of the general identity

$$\operatorname{Im} [g_\lambda,\overline{g}_\lambda](x) = -2\, |g_\lambda(x)|^2 \cdot \operatorname{Im} \frac{g_\lambda'(x)}{g_\lambda(x)} \,, \tag{6.14}$$

the integrand in (6.12) is negative under assumption (6.13). This explains the choice of the sign on the right hand side of (6.12).

Under condition (6.13) the limit (6.12) exists P – a.e. and is independent of the special choice of the solution g_λ. This can be seen as follows. A simple calculation with the Wronskian, which we postpone to the end of this section, shows

5) $$\left| \int_0^L \mathrm{Im} \frac{g_\lambda'(x)}{g_\lambda(x)} dx - \int_0^L \mathrm{Im} \frac{\tilde{g}_\lambda'(x)}{\tilde{g}_\lambda(x)} dx \right| < \pi \qquad , L > 0,$$

if $g_\lambda, \tilde{g}_\lambda$ are solutions of (1.3) satisfying (6.13). It is therefore sufficient to show that the limit (6.12) exists for one suitable g_λ satisfying (6.13). For this purpose we choose $g_\lambda = \varphi_\lambda - i\psi_\lambda$; then g_λ is real, i.e. the phase of g_λ is an integer valued multiple of π, if and only if ψ_λ has a zero, and, moreover, assumption (6.13) is satisfied. This suggests that the integral on the right hand side of (6.12) counts the number of zeroes of ψ_λ. Indeed,

$$\frac{d}{dx} \mathrm{Im} \log g_\lambda(x) = \mathrm{Im} \frac{g_\lambda'(x)}{g_\lambda(x)}$$

is strictly negative according to (6.14) and therefore one has exactly one zero of ψ_λ per decrease of Im log g_λ by π, i.e.

6) $$\left| - \int_0^L \mathrm{Im} \frac{g_\lambda'(x)}{g_\lambda(x)} dx - \pi \cdot \nu^{(L)}(\psi_\lambda) \right| \leq \pi \qquad \text{for } L \geq 0,$$

where $\nu^{(L)}(\psi_\lambda)$ denotes the number of zeroes of ψ_λ in the interval (0,L). It remains to show that the limit

$$\lim_{L \to \infty} \frac{1}{L} \nu^{(L)}(\psi_\lambda) \qquad , \lambda \in \mathbf{R},$$

does exist. In the case of a zero potential, the zeroes of $\psi_\lambda(x) = \frac{\sin(\sqrt{\lambda}x)}{\sqrt{\lambda}}$ on the positive half axis move to the left with increasing λ, and $\nu^{(L)}(\psi_\lambda)$ is related to the eigenvalues of $H^{(0,L)}$ by

$$\nu^{(L)}(\psi_\lambda) = \#\{j: \lambda_j^{(L)} < \lambda\} \qquad , \lambda \in \mathbf{R}.$$

It follows from Sturm's oscillation theory (which we recall at the beginnings of section 6.3.), that this relation remains true also for non-zero potentials. Furthermore, it is easy to see that the definition (6.6) of N is not affected if one replaces it by $\lim\limits_{L\to\infty} N^{(0,L)}$.

Hence it follows

$$\begin{aligned}\lim_{L\to\infty}\frac{1}{L}\,\nu^{(L)}(\psi_\lambda) &= \lim_{L\to\infty}\frac{1}{L}\,\#\{j: \lambda_j^{(L)} < \lambda\}\\ &= \lim_{L\to\infty} N^{(0,L)}(\lambda)\\ &= N(\lambda) \qquad\qquad , \lambda\in\mathbf{R},\end{aligned}$$

and because of (6.16)

$$\alpha(\lambda) = \pi \cdot N(\lambda) \qquad , \lambda\in\mathbf{R}. \tag{6.17}$$

Under assumption (6.13) the limit (6.12) therefore exists P – a.e. independently of the choice of g_λ, and the phase α is identified with $\pi \cdot N$.

We refer to α as the *rotation number* since it has the following geometric interpretation. For a given $\lambda\in\mathbf{R}$ let g_λ be a non-zero solution of (1.3), say $g_\lambda = \psi_\lambda$. It is convenient to use polar coordinates in the $g_\lambda' - g_\lambda$-plane and to write

$$\begin{aligned} g_\lambda'(x) &= r_\lambda(x)\cos\theta_\lambda(x)\\ g_\lambda(x) &= r_\lambda(x)\sin\theta_\lambda(x) \quad .\end{aligned} \tag{6.18}$$

Then x is a zero of g_λ if and only if $\theta_\lambda(x) \equiv 0 (\mathrm{mod}\ \pi)$ and

$$\alpha(\lambda) = \lim_{L\to\infty}\frac{\theta_\lambda(L)-\theta_\lambda(0)}{L} = \lim_{L\to\infty}\frac{1}{L}\int_0^L \frac{d}{dx}\theta_\lambda(x)\,dx \tag{6.19}$$

measures the average increase of the angle in the $g_\lambda' - g_\lambda$-plane.

Since the definition of the Ljapunov exponent γ was already given in (3.9) (alternatively, the Ljapunov exponent can be defined on the basis of (6.11) as will be seen in section 6.3. below), the basic notions we need are now defined. To summarize we have the following diagram:

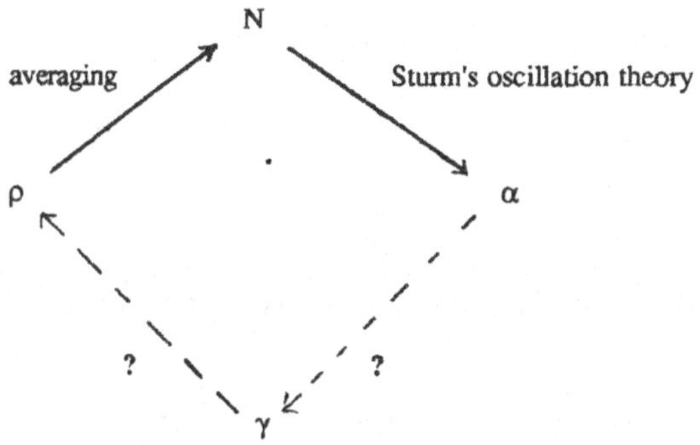

An important task will consist in finding a relation between α and γ. Whereas on the basis of the notions and ideas discussed so far a solution to this problem is by no means obvious, it is easy to give the answer to the question posed at the beginning of this section. To do so we formally differentiate (6.10) with respect to λ, use (6.17) and obtain

$$\pi \cdot E\left[\frac{d\rho_{ac}}{d\mu}\right] = \pi \cdot \frac{d}{d\lambda} N(\lambda) = \frac{d}{d\lambda}\alpha(\lambda) \qquad \mu - \text{a.e.}$$

(we do not care about the interchange of differentiation and expectation since later on no use of the above formula is made). Therefore the implication

0) $$\gamma(\lambda) = 0 \quad \text{for } \mu\text{-almost all } \lambda \text{ in a Borel set } A$$
$$\Rightarrow \quad \pi \cdot \frac{d\rho_{ac}}{d\mu}(\lambda;q) > 0 \qquad \text{for } \mu\otimes P\text{-almost all } (\lambda,q) \in A \times \Omega$$

plays the role of the pointwise analogue for individual potentials q of the implication

1) $$\gamma(\lambda) = 0 \quad \text{for } \mu\text{-almost all } \lambda \text{ in a Borel set } A$$
$$\Rightarrow \frac{d}{d\lambda}\alpha(\lambda) > 0 \qquad \text{for } \mu\text{-almost } \lambda \in A,$$

since (6.21) results from (6.20) by averaging with respect to P.

In the following sections we try to explain implication (6.20). Since the Ljapunov exponent γ is an averaged quantity, whereas the spectral measure itself is defined pointwise for individual potentials q, we have in particular to answer the following questions.

Question 1. Is there some analogue to Floquet theory for individual non-periodic potentials q?

Question 2. How is such a pointwise theory related to averaged quantities like α and γ?

Until now it was sufficient to consider equation (1.3) for real λ. In the following however it will be essential to work with complex numbers λ. In particular this will allow us to express the right hand side of (6.20) in terms of the Green's function G_λ of the operator H, where $\lambda \in \mathbf{C}_+ = \{z \in \mathbf{C}:\ \mathrm{Im}\, z > 0\}$. For $\lambda \in \mathbf{C}_+$ the Green's function $G_\lambda(x,y;q)$ $(x,y \in \mathbf{R})$ is defined as the kernel of the bounded operator $(H-\lambda)^{-1}$, i.e.

$$G_\lambda(x,y;q) = \int_{-\infty}^{+\infty} \frac{1}{\eta-\lambda}\, \mathcal{E}(d\eta;x,y) \qquad , \lambda \in \mathbf{C}_+ .$$

In particular, $G_\lambda(0,0;q)$ is the Stieltjes transform of the spectral measure $\rho(d\eta;q)$, since

$$(6.22) \qquad G_\lambda(0,0;q) = \int_{-\infty}^{+\infty} \frac{1}{\eta-\lambda}\, \mathcal{E}(d\eta;0,0) = \int_{-\infty}^{+\infty} \frac{\rho(d\eta;q)}{\eta-\lambda} \quad , \qquad \lambda \in \mathbf{C}_+ .$$

From

$$\mathrm{Im}\, G_{\xi+i\varepsilon}(0,0;q) = \int_{-\infty}^{+\infty} \frac{\varepsilon}{(\eta-\xi)^2+\varepsilon^2}\, \rho(d\eta;q) \qquad (\xi \in \mathbf{R}\ ,\ \varepsilon>0)$$

one easily obtains the absolutely continuous part of the spectral measure in terms of the boundary values of the Green's function by

23) $$\lim_{\varepsilon \downarrow 0} \frac{1}{\pi} \operatorname{Im} G_{\xi+i\varepsilon}(0,0;q) = \frac{1}{\pi} \operatorname{Im} G_{\xi+i\varepsilon}(0,0;q) = \frac{d\rho_{ac}}{d\mu}(\xi,q)$$

for μ-almost all $\xi \in \mathbf{R}$ (see for example section 9.4 in Richtmyer (1978)). Therefore, (6.20) can be written in the form

24) $$\gamma(\xi) = 0 \quad \text{for } \mu\text{-almost all } \xi \text{ in a Borel set A}$$
$$\Rightarrow \quad \operatorname{Im} G_{\xi+i0}(0,0;q) > 0 \quad \text{for } \mu \otimes P \text{ - almost all } (\xi,q) \in A \times \Omega.$$

Before we turn in the following sections to the proof of (6.24), we indicate the proof of (6.15) under assumption (6.13). We note that neither (6.13) nor (6.15) is affected if we replace g_λ by $a \cdot g_\lambda$ when $a \neq 0$ is constant. Therefore we may assume $g_\lambda(0) = \tilde{g}_\lambda(0) = 1$ so that because of (6.13) we have $\operatorname{Im} g_\lambda'(0) < 0$ and $\operatorname{Im} \tilde{g}_\lambda'(0) < 0$. If (6.15) were false, there would exist a number $L_0 > 0$ for which

$$\left| \int_0^{L_0} \left(\operatorname{Im} \frac{g_\lambda'(x)}{g_\lambda(x)} - \operatorname{Im} \frac{\tilde{g}_\lambda'(x)}{\tilde{g}_\lambda(x)} \right) dx \right| = \pi ,$$

and therefore there would exist a positive number c so that $g_\lambda + c \cdot \tilde{g}_\lambda = 0$ for $x = L_0$. On the other hand $h_\lambda = g_\lambda + c \cdot \tilde{g}_\lambda$ is a solution and satisfies $\operatorname{Im}[h_\lambda, \bar{h}_\lambda](0) > 0$. This follows from (6.14) because of

$$\operatorname{Im} \frac{h_\lambda'}{h_\lambda} \Big|_{x=0} = \frac{1}{1+c} (\operatorname{Im} g_\lambda' + \operatorname{Im} \tilde{g}_\lambda') \Big|_{x=0} < 0.$$

Therefore the solution h_λ does not have zeroes. This contradiction proves (6.15).

. *Heart of the proof*

We address the questions posed in the previous section and explain the main steps of the proof of (6.24) on a non-technical level. To do so we try to extend the ideas sketched in section 2.2. to non-periodic ergodic potentials. We proceed in three steps working first with individual q and then, in the last step, passing to averaged quantities.

Step 1. Weyl theory as a pointwise analogue of Floquet theory.

We consider equation (1.3) for a given $\lambda \in \mathbf{C}_+$. We may think of a solution g_λ as the result of multiple scattering on the potential q (cf. Figure 7). In contrast to example 2.2 we now cannot ask how solutions transform after one period. Instead we ask how solutions change after a step with infinitesimal width Δx and we look, so to say, for an infinitesimal analogue of Floquet solutions (cf. (2.21) and (2.22)). We look for exponents $m_\pm(\lambda;q) \in \mathbf{C}$ and solutions $f_\pm(\lambda;\cdot)$ of (1.3) satisfying

$$f_\pm(\lambda;x+\Delta x) \approx e^{\pm m_\pm(\lambda;\theta_x q)\cdot \Delta x} \cdot f_\pm(\lambda;x) \quad , x \in \mathbf{R},\ \Delta x \text{ small.} \tag{6.25}$$

If solutions of this special type exist, the real part $\operatorname{Re} m_\pm(\lambda;\theta_x q)\cdot \Delta x$ describes, roughly speaking, the infinitesimal change of the amplitude of the solution $f_\pm(\lambda;\cdot)$ after "scattering on the potential q at the place x", and the imaginary part $\operatorname{Im} m_\pm(\lambda;\theta_x q)\cdot \Delta x$ describes the infinitesimal change of the phase after "scattering at the place x".

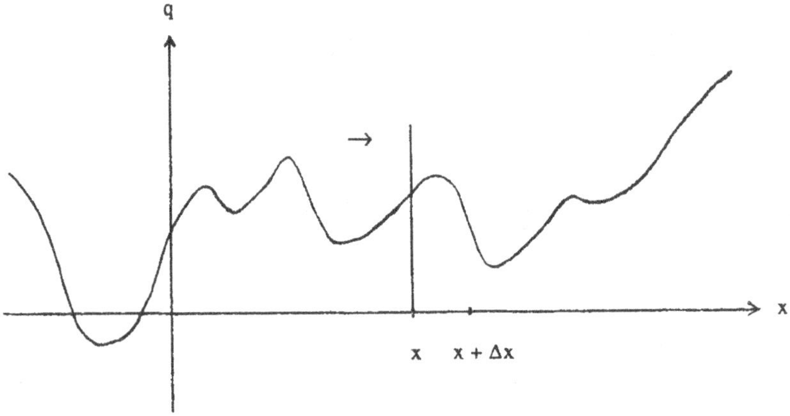

Figure 11

Such solutions do indeed exist. This is the content of a result by H. Weyl. To formulate it we denote the intervals $(0,+\infty)$ and $(-\infty,0)$ by $\mathbf{R}_+$ and $\mathbf{R}_-$ respectively.

Proposition 3. *For* $\lambda \in \mathbf{C}_+$ *there exist two linearly independent solutions* $f_\pm(\lambda;\cdot)$ *of* (1.3) *such that* $f_\pm(\lambda;\cdot)$, *restricted to* $\mathbf{R}_\pm$, *belongs to* $L^2(\mathbf{R}_\pm)$. *For* $q \in \Omega$, *the solutions* $f_\pm(\lambda;\cdot)$ *are unique up to a factor. There exist functions* $m_\pm: \mathbf{C}_+ \times \Omega \to \mathbf{C}_+$, *the so called Titchmarsh-Weyl functions, such that*

.6) $$\frac{\frac{d}{dx} f_\pm(\lambda;x)}{f_\pm(\lambda;x)} = \pm\, m_\pm(\lambda;\theta_x q) \qquad , x \in \mathbf{R}.$$

For $q \in \Omega$ *the functions* $\lambda \mapsto m_\pm(\lambda;q)$ *are holomorphic on* $\mathbf{C}_+$ *with positive imaginary parts.*

Following Weyl (1950) we will explain the basic ideas of the proof of Proposition 3 in section 6.3. below. Relation (6.26) is the precise formulation of (6.25). From (6.26) and from $\frac{d^2}{dx^2} f_\pm(\lambda;x) = (q(x) - \lambda)\cdot f_\pm(\lambda;x)$ it immediately follows, that the functions $x \mapsto m_\pm(\lambda;\theta_x q)$ satisfy the Riccati equations

7) $$\pm \frac{d}{dx} m_\pm(\lambda;\theta_x q) = q(x) - \lambda - (m_\pm(\lambda;\theta_x q))^2 .$$

Later on, the proof of some fundamental analytical identities will be based on these Riccati equations. An example for such an application will be given in the following two steps.

Step 2. Deduction of a relation between the real and the imaginary part of $m_\pm$ in an appropriate divergence form.

With Figure 5 in mind we try to find a relation between the functions $x \mapsto \mathrm{Re}\, m_\pm(\lambda;\theta_x q)$ and $x \mapsto \mathrm{Im}\, m_\pm(\lambda;\theta_x q)$. For this purpose we take the imaginary part on both sides of (6.27) and obtain

8) $$\pm \frac{d}{dx} \mathrm{Im}\, m_\pm(\lambda;\theta_x q) = -\mathrm{Im}\lambda - 2(\mathrm{Re}\, m_\pm(\lambda;\theta_x q))\cdot(\mathrm{Im}\, m_\pm(\lambda;\theta_x q)) .$$

From Proposition 3 we know $\mathrm{Im}\, m_\pm(\lambda;\theta_x q) > 0$. Therefore we can divide both sides of (6.28) by $\mathrm{Im}\, m_\pm(\lambda;\theta_x q)$ and get for $\lambda \in \mathbf{C}_+$

9) $$\left(\frac{\mathrm{Im}\lambda}{\mathrm{Im}\, m_\pm(\lambda;\theta_x q)} + 2\,\mathrm{Re}\, m_\pm(\lambda;\theta_x q)\right) \pm \frac{d}{dx} \log(\mathrm{Im}\, m_\pm(\lambda;\theta_x q)) = 0 .$$

It is important that the last summand in (6.29) is in divergence form. We take advantage of this fact in the next step.

Step 3. Taking expectations in (6.29) and passing to the limit $\mathrm{Im}\,\lambda \downarrow 0$.
Now we use the stationarity of the potential. If we take expectations in equation (6.29), the divergence term vanishes and we obtain

$$(6.30) \qquad -\frac{2\,\mathrm{Re}\,E[m_\pm(\lambda)]}{\mathrm{Im}\lambda} = E\left[\frac{1}{\mathrm{Im}\,m_\pm(\lambda)}\right] \qquad , \lambda \in \mathbf{C}_+ .$$

According to the heuristic considerations given above in step 1, $E[m_\pm(\lambda)]$ describes, roughly speaking, the change of the function $f_\pm(\lambda, \cdot\,)$ after repeated scattering *in the average*. This suggests a connection with the generalized Floquet exponent $w(\lambda)$, which we will define later also for $\lambda \in \mathbf{C}_+$ along the lines of the Ansatz (6.11). In fact, in section 6.3. we will see that the Floquet exponent, generalized in this way, just coincides with the average of the Titchmarsh-Weyl functions

$$(6.31) \qquad w(\lambda) = E[m_+(\lambda)] = E[m_-(\lambda)] \qquad , \lambda \in \mathbf{C}_+ .$$

Moreover, w is holomorphic in the upper half complex plane and is related to the Ljapunov exponent and to the rotation number by

$$(6.32) \qquad \gamma(\xi) = -\lim_{\varepsilon \downarrow 0} \mathrm{Re}\, w(\xi + i\varepsilon) = -\,\mathrm{Re}\;w(\xi + i0) \qquad , \xi \in \mathbf{R} ,$$

and

$$(6.33) \qquad \alpha(\xi) = \lim_{\varepsilon \downarrow 0} \mathrm{Im}\, w(\xi + i\varepsilon) = \mathrm{Im}\;w(\xi + i0) \qquad , \xi \in \mathbf{R} .$$

Taking this for granted, the proof of (6.24) can easily be finished on a heuristic level as follows. Let $A \subset \mathbf{R}^1$ be a bounded Borel set such that

$$(6.34) \qquad \gamma(\xi) = 0 \qquad \text{for } \mu\text{-almost all } \xi \in A.$$

Integrating the left hand side of (6.30) on the set A and using (6.33) and (6.34), one obtains

$$(6.35)\qquad \lim_{\varepsilon\downarrow 0}\int_A \frac{-\operatorname{Re} E[m_\pm(\xi+i\varepsilon)]}{\varepsilon}\,\mu(d\xi)$$
$$= \lim_{\varepsilon\downarrow 0}\int_A \left(- \frac{\operatorname{Re} w(\xi+i\varepsilon) - \operatorname{Re} w(\xi+i0)}{\varepsilon}\right)\mu(d\xi)\,.$$

Assume that the limit and the integral in the last line can be interchanged. Then the Cauchy-Riemann equations for the holomorphic function w and (6.33) suggest

$$(6.36)\qquad \lim_{\varepsilon\downarrow 0}\int_A - \frac{\operatorname{Re} E[m_\pm(\xi+i\varepsilon)]}{\varepsilon}\,\mu(d\xi)$$
$$= \int_A \lim_{\varepsilon\downarrow 0}\left(- \operatorname{Re}\frac{w(\xi+i\varepsilon) - w(\xi+i0)}{\varepsilon}\right)\mu(d\xi)$$
$$= \int_A \frac{d}{d\xi}\operatorname{Im} w(\xi+i0)\ \ \mu(d\xi)$$
$$= \int_A \frac{d}{d\xi}\alpha(\xi)\ \ \mu(d\xi)$$
$$= \pi\cdot\int_A \frac{d}{d\xi}N(\xi)\ \ \mu(d\xi)\ \le\ \pi\cdot\int_A N(d\xi) < \infty\ .$$

In section 6.4. we will see that the heuristic consideration (6.36) is indeed correct.

Using Fatou's lemma one obtains from (6.30) and from (6.36)

$$E\int_A \liminf_{\varepsilon\downarrow 0}\frac{1}{\operatorname{Im} m_\pm(\xi+i\varepsilon;q)}\,\mu(d\xi) < \infty \qquad \text{and hence}$$

(6.37) $$\operatorname{Im} m_{\pm}(\xi + i0;q) > 0 \text{ for } \mu \otimes P - \text{almost all } (\xi,q) \in A \times \Omega.$$

Notice that for μ-almost all $\xi \in \mathbf{R}$ the limit $m_{\pm}(\xi + i0;q) = \lim_{\varepsilon \downarrow 0} m_{\pm}(\xi + i\varepsilon;q)$ exists and is finite; this is a general property of Herglotz functions, i.e. of holomorphic functions $h: \mathbf{C}_+ \to \mathbf{C}_+$ (see for example § 7 in Kotani (1987)). In order to see that (6.37) implies (6.24), it only remains to express $G_\lambda(0,0;q)$ by means of $m_{\pm}(\lambda;q)$ $(\lambda \in \mathbf{C}_+)$. Analogously to (2.31) one has

(6.38) $$G_\lambda(x,y;q) = \frac{f_-(\lambda;x)\, f_+(\lambda;y)}{[f_+(\lambda)\, ,\, f_-(\lambda)]} \quad , \; x \le y ,$$

and hence by (6.26)

(6.39) $$G_\lambda(0,0;q) = \frac{f_-(\lambda;0)\, f_+(\lambda;0)}{[f_+(\lambda)\, ,\, f_-(\lambda)]}$$

$$= - \frac{1}{m_-(\lambda;q) + m_+(\lambda;q)} \quad , \; \lambda \in \mathbf{C}_+ .$$

From (6.37) and (6.39) one concludes

$$\operatorname{Im} G_{\xi+i0}(0,0;q) = \operatorname{Im} \left(- \frac{1}{m_-(\xi + i0;q) + m_+(\xi + i0;q)} \right)$$

$$= \frac{\operatorname{Im} m_-(\xi + i0;q) + \operatorname{Im} m_+(\xi + i0;q)}{|\, m_-(\xi + i0;q) + m_+(\xi + i0;q)\, |^2} > 0$$

for $\mu \otimes P$ - almost all $(\xi,q) \in A \times \Omega$. This proves (6.24).

To summarize, the above proof is essentially based on three main ideas:

(i) Weyl's spectral theory of singular Sturm-Liouville operators is related to a generalized Floquet theory for ergodic potentials which, roughly speaking, results from Weyl's theory by averaging. This part of the proof, which is of general conceptual importance, will be explained in section 6.3..

(ii) The Titchmarsh-Weyl functions and the generalized Floquet exponent are holomorphic in the complex upper half plane. Their real and imaginary parts are therefore connected by the Cauchy-Riemann equations, from which some kind of dispersion relation between amplitude and phase is obtained. Such a relation is also behind Figure 5, which was the starting point for the proof sketched above.

(iii) There exists a relation between the real and the imaginary part of the Titchmarsh-Weyl functions, which is in an appropriate divergence form. In the theory to be developed below, equations in such a form play a role similar to conservation laws; by taking expectations, the divergence term vanishes and a relation between averaged quantities can be obtained in this way. This method will be applied repeatedly in the following section.

6.3. *Introduction to an extended Floquet-Weyl theory*

Now we try to make precise the concepts which were previously introduced on a heuristic level. After recalling Sturm's oscillation theory in subsection a., the basic idea of the proof of Proposition 3 will be explained in subsection b., and in the following subsection an extension of the notion of Floquet exponent to the case of general ergodic potentials is given. In subsection d. the connection between these ideas and Hadamard's factorization theorem for entire functions is described and an important relation between N and γ, the so called Thouless formula, is derived.

It is shown, how Sturm's oscillation theory, Weyl's theory of singular Sturm-Liouville operators, Floquet theory, and Hadamard's factorization theorem tie together on the basis of appropriate averaging procedures. The contents of this section may be summarized in the following diagram.

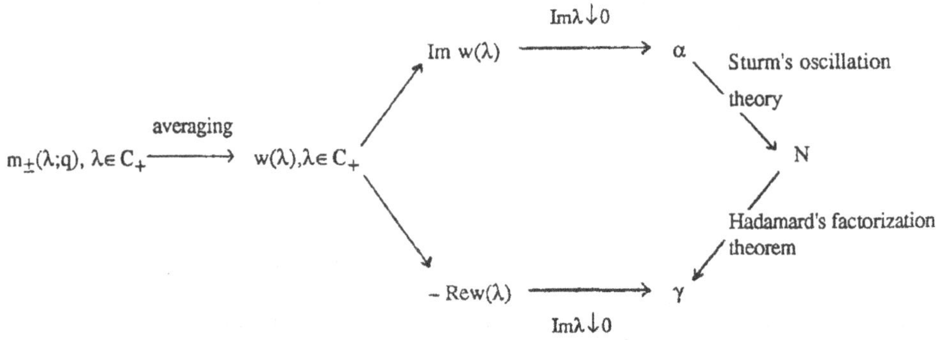

a. Sturm's oscillation theory

The spectral theory of Sturm–Liouville operators originates with the following basic observation. Let $Q\colon \mathbf{R} \to \mathbf{R}$ be a continuous function. We consider a non-zero solution $h = h_Q$ of the equation

$$
\begin{aligned}
&h''(x) = Q(x) \cdot h(x) \;, \quad -\infty < x < +\infty \\
&h(0) = 0
\end{aligned}
\tag{6.40}
$$

and, with the case $Q \equiv$ constant in mind, we ask what the oscillatory behaviour of the solution h_Q is, when Q is varying. What we will show is that the zeroes of h_Q on the positive half axis move to the left with decreasing Q. More precisely, let Q_1 and Q_2 be continuous functions on **R** with $Q_1(x) \le Q_2(x)$, $x \in \mathbf{R}$. We denote the corresponding solutions of (6.40) by $h_1 = h_{Q_1}$ and by $h_2 = h_{Q_2}$ respectively and we consider two consecutive zeroes a and b of h_2, $0 \le a < b$. For definiteness we assume

$$h_2(a) = h_2(b) = 0 \text{ and } h_2(x) > 0 \quad \text{for } a < x < b, \text{ hence } h_2'(a) > 0,\ h_2'(b) < 0.$$

Then the solution h_1 has a zero in the interval (a,b), provided that Q_1 and Q_2 are not identical on (a,b) (If Q_1 and Q_2 are identical on (a,b) and if h_1 has no zeroes on (a,b), then $h_1(a) = h_1(b) = 0$). To prove this we assume the contrary, say $h_1(x) > 0$ for $x \in (a,b)$. From Green's formula we obtain

41) $$[h_1,h_2]\,(x)\,\Big|_a^b = \int_a^b (h_1(x)\,h_2''(x) - h_2(x)\,h_1''(x))\,dx$$

$$= \int_a^b (Q_2(x) - Q_1(x))\,h_1(x)h_2(x)\,dx > 0.$$

Since the left hand side of (6.41) is $h_1(b) \cdot h_2'(b) - h_1(a) \cdot h_2'(a) \le 0$, we have a contradiction.

To apply this result with $Q = q - \lambda$ ($q \in \Omega$, $\lambda \in \mathbf{R}$), we denote by $h_Q = \psi_\lambda^q$ that solution of (6.40) which is normalized by $h_Q'(0) = 1$, and we consider for $L > 0$ the operator $H^{(0,L)} = -\frac{d^2}{dx^2} + q$ in the interval (0,L) with Dirichlet boundary conditions. If the eigenvalues $\lambda_j^{(L)}$ of $H^{(0,L)}$ ($j \ge 1$) are numbered according to their order, the jth eigenfunction is just $\psi^q_{\lambda_j^{(L)}}$. From Sturm's observation the following consequences can be drawn.

(i) The jth eigenfunction has exactly $j-1$ zeroes in (0,L). This nodal theorem follows, since the jth eigenfunction has at least $j-1$ zeroes in (0,L) according to Sturm's theory, and it has at most $j-1$ zeroes because of the general fact that the zeroes of the jth eigenfunction divide the domain (0,L) in at most j domains free of zeroes, a result which is true in arbitrary dimension and which can be proved for example by means of the maximum-minimum principle (see Chapter VI, § 6 in Courant and Hilbert (1953)).

(ii) Denoting by $\nu^{(L)}(\psi_\lambda^q)$ = the number of zeroes of ψ_λ^q in (0,L), the integrated density of states N_q corresponding to the potential q is given by

(6.42) $$N_q(\lambda) = \lim_{L\to\infty} \frac{1}{L} \nu^{(L)}(\psi_\lambda^q) \qquad , \lambda \in \mathbf{R}.$$

This follows from (i) and from Sturm's theory, since for λ with $\lambda_n^{(L)} < \lambda \le \lambda_{n+1}^{(L)}$ ($n \in \mathbf{N}$) one has $\#\{j: \lambda_j^{(L)} < \lambda\} = n = \nu^{(L)}(\psi^q_{\lambda^{(L)}_{n+1}}) = \nu^{(L)}(\psi_\lambda^q)$.

(iii) N_q is monotonically decreasing in q:

(6.43) $$q_1 \le q_2 \Rightarrow N_{q_1}(\lambda) \ge N_{q_2}(\lambda) \qquad , \lambda \in \mathbf{R},$$

because an application of Sturm's result with $Q_i = q_i - \lambda$ $(i = 1,2)$ shows $\nu^{(L)}(\psi_\lambda^{q_1}) \ge \nu^{(L)}(\psi_\lambda^{q_2})$.

b. Weyl's theory of singular Sturm - Liouville operators

We consider equation (1.3) for $\lambda \in \mathbf{C}_+$ and denote as before the fundamental matrix by

$$Y_\lambda(x;q) = \begin{pmatrix} \varphi_\lambda(x) & \psi_\lambda(x) \\ \varphi'_\lambda(x) & \psi'_\lambda(x) \end{pmatrix} .$$

If $f_\pm(\lambda)$ is a solution of (1.3) which satisfies (6.26) and which is normalized by $f_\pm(\lambda;0) = 1$, the Titchmarsh-Weyl functions are given by

(6.44) $$\pm\, m_\pm(\lambda;q) = \frac{f'_\pm(\lambda)}{f_\pm(\lambda)}\Big|_{x=0} = f'_\pm(\lambda;0)$$

and $f_\pm(\lambda)$ is of the form

(6.45) $$f_\pm(\lambda) = \varphi_\lambda \pm m_\pm(\lambda;q)\psi_\lambda \,.$$

Conversely, in order to prove Proposition 3 we have to find coefficients $m_\pm(\lambda;q) \in \mathbb{C}$ such that $f_\pm(\lambda)|_{R_\pm} \in L^2(R_\pm)$, if the functions $f_\pm(\lambda)$ are defined by (6.45) with these coefficients $m_\pm(\lambda;q)$.

The construction of $m_\pm(\lambda;q)$ is based, once more, on an appropriate use of Green's formula, which for solutions g_λ of (1.3) yields

(6.46) $$[g_\lambda, \bar{g}_\lambda]\Big|_0^L = \int_0^L (g_\lambda(x)\bar{g}_\lambda''(x) - \bar{g}_\lambda(x)g_\lambda''(x))\,dx$$

$$= 2i(\mathrm{Im}\lambda) \cdot \int_0^L |(g_\lambda(x)|^2\,dx \qquad , L > 0.$$

If we choose g_λ so that the left hand side of (6.46) is bounded in L, the integral on the right hand side of (6.46) remains bounded as $L \to \infty$ and we can hope to find in this way a solution g_λ which belongs to $L^2(R_+)$. The left hand side of (6.46) does not depend on the values of g_λ and of g_λ' at L, if $[g_\lambda, \bar{g}_\lambda]$ (L) = 0, i.e. if the logarithmic derivative $\frac{g_\lambda'}{g_\lambda}\Big|_{x=L}$ is *real* at the boundary L. In the case of the boundary $-L$ one can argue similarly.

We first investigate the geometrical meaning of this boundary condition, which is imposed on the left hand side of (6.46), and then translate it into analytic terms by means of the right hand side of (6.46).

More precisely, we look for solutions

(6.47) $$g_\pm^{(L)}(\lambda) = \varphi_\lambda \pm m_\pm^{(L)}(\lambda;q) \cdot \psi_\lambda$$

with coefficients $m_\pm^{(L)}(\lambda;q) \in \mathbb{C}$ to be determined so that

(6.48) $$\frac{\frac{d}{dx} g_\pm^{(L)}(\lambda;x)}{g_\pm^{(L)}(\lambda;x)}\Bigg|_{\pm L} = \frac{\varphi_\lambda'(\pm L) \pm m_\pm^{(L)}(\lambda;q) \cdot \psi_\lambda'(\pm L)}{\varphi_\lambda(\pm L) \pm m_\pm^{(L)}(\lambda;q) \cdot \psi_\lambda(\pm L)}$$

is *real*, which means that $m_\pm^{(L)}(\lambda;q)$ has to belong to the set

(6.49) $$S_{\pm}^{(L)} = \{ z \in \mathbb{C} : \ \frac{a+bz}{c+dz} = \frac{\bar{a}+\bar{b}\bar{z}}{\bar{c}+\bar{d}\bar{z}} \} ,$$

where for short we have set $\varphi_{\lambda}'(\pm L) = a$, $\pm\ \psi_{\lambda}'(\pm L) = b$, $\varphi_{\lambda}(\pm L) = c$, $\pm\ \psi_{\lambda}(\pm L) = d$. By cross multiplication one calculates directly that $S_{\pm}^{(L)}$ is a circle in the complex plane, whose radius is given by

(6.50) $$r_{\pm}^{(L)} = \frac{|bc-ad|}{|b\bar{d}-d\bar{b}|} = \frac{|[\varphi_{\lambda},\psi_{\lambda}]\ (\pm L)|}{|[\psi_{\lambda},\overline{\psi_{\lambda}}]\ (\pm L)|} = \frac{1}{|[\psi_{\lambda},\overline{\psi_{\lambda}}]\ (\pm L)|} .$$

On the other hand, under the condition $[g_{\pm}^{(L)},\bar{g}_{\pm}^{(L)}]\ (L) = 0$ equation (6.46) reduces to

$$2i \cdot \mathrm{Im}(m_{+}^{(L)}\ (\lambda;q)) = 2i\ (\mathrm{Im}\lambda) \int_0^L |\varphi_{\lambda}(x) + m_{+}^{(L)}(\lambda;q)\ \psi_{\lambda}(x)|^2 dx.$$

Therefore, the circle $S_{+}^{(L)}$ can also be described by

(6.51) $$S_{+}^{(L)} = \{ z \in \mathbb{C}_{+} : \int_0^L |\varphi_{\lambda}(x) + z\psi_{\lambda}(x)|^2\ dx = \frac{\mathrm{Im}\, z}{\mathrm{Im}\, \lambda} \},$$

and one recognizes $S_{+}^{(L)}$ as the boundary of the disc

$$D_{+}^{(L)} = \{\ z \in \mathbb{C}_{+} : \int_0^L |\varphi_{\lambda}(x) + z\psi_{\lambda}(x)|^2\ dx \ \le \frac{\mathrm{Im}\, z}{\mathrm{Im}\, \lambda} \} .$$

From this one can see that the discs $D_{+}^{(L)}$ are decreasing and hence the circles $S_{+}^{(L)}$ are shrinking, if L is increased, and similarly the circles $S_{-}^{(L)}$. Their radius is given by

(6.52) $$r_{\pm}^{(L)} = (\pm 2(\mathrm{Im}\lambda) \int_0^{\pm L} |\psi_{\lambda}(x)|^2\ dx\ )^{-1},$$

as follows from (6.50) and from formula (6.46), applied with $g_\lambda = \psi_\lambda$.

If one can show that in the limit $L \to \infty$ the circles $S_\pm^{(L)}$ *shrink to a point* $m_\pm(\lambda;q) \in C_+$, one expects that $f_\pm(\lambda)\big|_{R_\pm}$, defined by (6.45) with this coefficient $m_\pm(\lambda;q)$, belongs to $L^2(R_\pm)$.

This is indeed correct. Since q is bounded from below, one is in the so called limit point case and the following holds true (for the proof we refer for example to section 10.9 in Richtmyer (1978) and the references given there): In the limit $L \to \infty$, the circles $S_\pm^{(L)}$ shrink to a point $m_\pm(\lambda;q)$; $\lambda \mapsto m_\pm(\lambda;q)$ is a holomorphic function on C_+ with positive imaginary part; the function $f_\pm(\lambda) = \varphi_\lambda \pm m_\pm(\lambda;q)\psi_\lambda$ belongs to $L^2(R_\pm)$ and its norm is related to Im $m_\pm(\lambda)$ by

(6.53) $$\int_{R_\pm} |f_\pm(\lambda;x)|^2 dx = \frac{\text{Im } m_\pm(\lambda)}{\text{Im}\lambda} \quad , \lambda \in C_+ .$$

Given this result, the proof of the remaining part of Proposition 3 is easy. Obviously, the functions $f_+(\lambda)$ and $f_-(\lambda)$ are linearly independent, since otherwise $f_\pm(\lambda)$ would be square integrable on the real line and λ would be a non-real eigenvalue. Furthermore, $f_\pm(\lambda)$ is, up to a constant factor, the unique solution of (1.3) which satisfies $f_\pm(\lambda)\big|_{R_\pm} \in L^2(R_\pm)$. This follows, since in the limit point case one has $\int_{R_\pm} |\psi_\lambda(x)|^2 dx = \infty$ because of (6.52), and hence $\varphi_\lambda + c \cdot \psi_\lambda \in L^2(R_\pm)$ if and only if $c = \pm m_\pm(\lambda;q)$. To prove (6.26) for $x \in \mathbf{R}$, we notice that $f_\pm(\lambda)$ has no zeroes (which will be verified in a moment), and hence we can define functions $h_\pm(\lambda;\cdot)$ by

$$h_\pm(\lambda;y) = \frac{f_\pm(\lambda;x+y)}{f_\pm(\lambda;x)} \quad , y \in \mathbf{R}.$$

$h_\pm(\lambda)$ is normalized by $h_\pm(\lambda;0) = 1$, solves the equation

$$\frac{d^2}{dy^2} h_\pm(\lambda;y) = (\theta_x q(y) - \lambda)\, h_\pm(\lambda;y) \qquad y \in \mathbf{R},$$

and belongs to $L^2(\mathbf{R}_\pm)$. Such a solution is unique. Applying the theory outlined above to the potential $\theta_x q$, we therefore obtain from (6.44)

$$\pm\, m_\pm(\lambda;\theta_x q) = \left.\frac{\frac{d}{dy} h_\pm(\lambda)}{h_\pm(\lambda)}\right|_{y=0} = \frac{\frac{d}{dx} f_\pm(\lambda;x)}{f_\pm(\lambda;x)} .$$

This proves (6.26).

It remains to verify that $f_\pm(\lambda)$ has no zeroes. From (6.46) one has

$$[f_+(\lambda)\,,\overline{f_+(\lambda)}\,](x)\Big|_0^L = 2i(\mathrm{Im}\,\lambda)\cdot\int_0^L |f_+(\lambda;x)|^2\,dx \qquad , L \geq 0,$$

hence by (6.53)

$$\lim_{L\to\infty}\,[f_+(\lambda),\overline{f_+(\lambda)}\,](L) = 2i(\mathrm{Im}\lambda)\cdot\int_0^\infty |f_+(\lambda;x)|^2\,dx + [f_+(\lambda)\,,\overline{f_+(\lambda)}]\,(0)$$

$$= 2i(\mathrm{Im}\lambda)\cdot\int_0^\infty |f_+(\lambda;x)|^2\,dx - 2i(\mathrm{Im}\;m_+(\lambda))$$

$$= 0.$$

From this one obtains for $a \in \mathbf{R}$

(6.54) $$\mathrm{Im}[f_+(\lambda),\,\overline{f_+(\lambda)}\,]\,(a) = -\,\mathrm{Im}[f_+(\lambda)\,,\,\overline{f_+(\lambda)}\,]\,(x)\Big|_a^\infty$$

$$= -\,2\cdot(\mathrm{Im}\,\lambda)\cdot\int_a^\infty |f_+(\lambda;x)|^2\,dx < 0 \;\;,$$

which proves $f_+(\lambda) \neq 0$. $f_-(\lambda) \neq 0$ can be shown similarly.

We conclude with two

Remarks. (i) Let $\lambda \in \mathbf{C}_+$. According to Weyl's theory, one has $f_\pm(\lambda)|_{\mathbf{R}_\pm} \in L^2(\mathbf{R}_\pm)$ pointwise for individual potentials. On the other hand it is easily seen that $|f_\pm(\lambda;x)|$ decreases exponentially as $x \to \pm\infty$ for P-almost all $q \in \Omega$. This follows, because (6.30) implies $\mathrm{Re}\, E[m_\pm(\lambda)] < 0$ and hence

$$f_\pm(\lambda;x) = f_\pm(\lambda;0) \cdot \exp\left(x \cdot \frac{1}{x}\int_0^x \pm m_\pm(\lambda;\theta_y q)dy\right)$$

with

$$\lim_{|x|\to\infty} \frac{1}{x}\int_0^x \pm \mathrm{Re}\, m_\pm(\lambda;\theta_y q)dy = \pm\, \mathrm{Re}\ E[m_\pm(\lambda)] = \mp|\,\mathrm{Re}\, E\,[m_\pm(\lambda)]\,|$$

by the pointwise ergodic theorem. The graph of $x \mapsto |f_\pm(\lambda;x)|$ looks as follows.

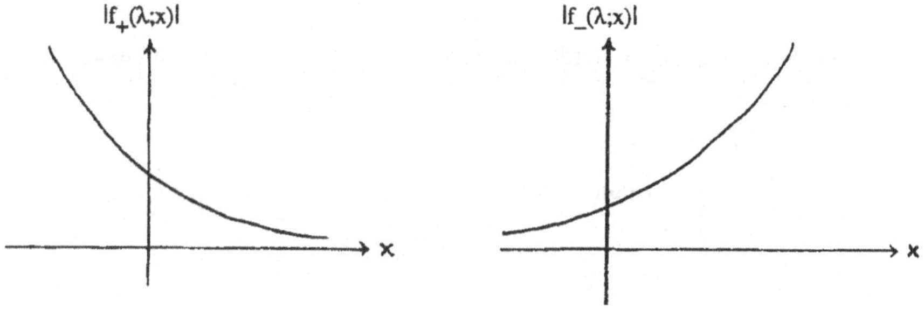

Figure 12

(ii) Equation (6.53) can be written in the form

5) $$\int_0^\infty \exp\left(2 \cdot \int_0^x \mathrm{Re}\, m_+(\lambda;\theta_y q)dy\right)dx = \frac{\mathrm{Im}\ m_+(\lambda)}{\mathrm{Im}\lambda}, \quad \lambda \in \mathbf{C}_+ .$$

This is a useful relation between $\mathrm{Re}\ m_+(\lambda)$ and $\mathrm{Im}\ m_+(\lambda)$, from which one easily obtains the following sharpening of implication (6.21). Let $A \subset \mathbf{R}^1$ be a Borel set. Then

(6.56) $$\gamma(\xi) = 0 \quad \text{for } \mu\text{-almost all } \xi \in A \Rightarrow \frac{d}{d\xi}(\alpha(\xi))^2 \geq 1 \quad \text{for } \mu\text{-almost all } \xi \in A.$$

In the case of a zero potential one has $(\alpha(\xi))^2 = \xi$ if $\gamma(\xi) = 0$, and equality holds on the right hand side of implication (6.56). For the proof of (6.56) one takes expectations on both sides of (6.55), applies Jensen's inequality and obtains

$$\begin{aligned}
\frac{\operatorname{Im} E[m_+(\lambda)]}{\operatorname{Im} \lambda} &= \int_0^\infty E[\exp(2\int_0^x \operatorname{Re} E[m_+(\lambda;\theta_y q)]\, dy)]dx \\
&\geq \int_0^\infty \exp(2\int_0^x \operatorname{Re} E[m_+(\lambda;\theta_y q)]dy\,)\, dx \\
&= -\frac{1}{2\operatorname{Re} E[m_+(\lambda)]} \qquad , \lambda \in C_+ .
\end{aligned}$$

Writing $\lambda = \xi + i\varepsilon$, integrating over a compact subset $K \subset A$ and using (6.31) one deduces

$$2\int_K \frac{-\operatorname{Re} w(\xi + i\varepsilon)}{\varepsilon} \cdot \operatorname{Im} w(\xi + i\varepsilon)\, \mu(d\xi) \geq \int_K 1\, \mu(d\xi),$$

from which one obtains in the limit $\varepsilon \downarrow 0$ by means of (6.36), (6.32) and (6.33)

$$2\int_K (\frac{d}{d\xi}\alpha(\xi)) \cdot \alpha(\xi)\ \mu(d\xi) \geq \int_K 1\, \mu(d\xi) \qquad , K \subset A.$$

(In section 6.4. it will be shown that the limit and the integral are interchangeable). This proves (6.56).

c. Extension of the notion of Floquet exponent

In order to define a generalized Floquet exponent in the case $\lambda \in C_+$, we proceed similarly as before in the definition of the phase α for real λ (see equation (6.11) in section 6.1.) and show

Proposition 4. *Let* $\lambda \in C_+$ *and* $g_\lambda(x;q)$ *be a solution of (1.3) which is measurable on* $\mathbf{R} \times \Omega$. *If* g_λ *satisfies*

$$\mathrm{Im}[g_\lambda, \bar{g}_\lambda](0) > 0, \tag{7}$$

then the limit

$$w(\lambda) = -\lim_{L\to\infty} \frac{1}{L} \int_0^L \frac{g_\lambda'(x)}{g_\lambda(x)}\, dx \tag{8}$$

exists P–a.e. *and is given by*

$$w(\lambda) = E[m_-(\lambda)] = E[m_+(\lambda)]. \tag{9}$$

Furthermore, w *is holomorphic on* C_+ *and has a positive imaginary part.*

Proof. We proceed in three steps.

Step 1. Proof of (6.58) in the case $g_\lambda = f_-(\lambda)$.

Condition (6.57) is satisfied for $f_-(\lambda)$ since

$\mathrm{Im}\,[f_-(\lambda), \overline{f_-(\lambda)}]\,(0) = 2\,|f_-(\lambda;0)|^2 \cdot \mathrm{Im}\, m_-(\lambda) > 0$. From (6.26) one obtains by means of the pointwise ergodic theorem

$$-\lim_{L\to\infty} \frac{1}{L} \int_0^L \frac{f_-'(\lambda;x)}{f_-(\lambda;x)}\, dx = \lim_{L\to\infty} \frac{1}{L} \int_0^L m_-(\lambda;\theta_x q)\, dx = E[m_-(\lambda)] \qquad \text{P–a.e..}$$

It is a minor technical point to verify that $m_\pm(\lambda)$ is integrable with respect to P. For the proof we refer to Lemma 1.3 in Kotani (1987) or to Lemma III.1.17 in Carmona and Lacroix (1990). From this Lemma and from its proof it also follows, that $E[m_\pm(\lambda)]$ can be differentiated with respect to λ by interchanging the differentiation with the expectation, a fact, which we will freely use in the following. Thus we conclude that w is a holomorphic function on C_+ with positive imaginary part, since $m_\pm(\cdot\,;q)$ has these properties.

Step 2. Reduction to the case $g_\lambda = f_-(\lambda)$.

Let g_λ be a solution of (1.3) satisfying (6.57). Since λ is complex, $\bar{g}_\lambda$ is no longer a solution and hence $[g_\lambda,\bar{g}_\lambda](x)$ is not constant. However, $\mathrm{Im}[g_\lambda,\bar{g}_\lambda](x)$ is increasing because of

$$\frac{d}{dx}\,\mathrm{Im}[g_\lambda,\bar{g}_\lambda](x) = 2(\mathrm{Im}\lambda)\cdot|g_\lambda(x)|^2 \qquad , x\in \mathbf{R},$$

which is the differentiated version of Green's formula (6.46). Hence $\mathrm{Im}\,[g_\lambda,\bar{g}_\lambda]\,(x) > 0$ for $x \geq 0$ and $g_\lambda(x) \neq 0$ for $x \geq 0$. Thus, the integrand on the right hand side of (6.58) is well defined under condition (6.57). Since the proof of (6.15) works for complex λ too, one has

$$\left|\int_0^L \mathrm{Im}\frac{g_\lambda'(x)}{g_\lambda(x)}\,dx - \int_0^L \mathrm{Im}\frac{f_-'(\lambda;x)}{f_-(\lambda;x)}\,dx\right| < \pi \qquad , L \geq 0,$$

for solutions g_λ satisfying (6.57), hence by step 1

$$\lim_{L\to\infty} \frac{1}{L}\int_0^L \mathrm{Im}\frac{g_\lambda'(x)}{g_\lambda(x)}\,dx = \lim_{L\to\infty} \frac{1}{L}\int_0^L \mathrm{Im}\frac{f_-'(\lambda;x)}{f_-(\lambda;x)}\,dx = \mathrm{Im}\,E[m_-(\lambda)\ \text{P–a.e..}$$

It remains to show

$$\lim_{L\to\infty} \frac{1}{L} \left| \int_0^L \mathrm{Re}\, \frac{g_\lambda'(x)}{g_\lambda(x)}\, dx \; - \int_0^L \mathrm{Re} \frac{f_-'(\lambda;x)}{f_-(\lambda;x)}\, dx \right| = 0 \qquad \text{P–a.e.,}$$

i.e.

$$\lim_{L\to\infty} \frac{1}{L} \left| \log \frac{|g_\lambda(L)|}{|g_\lambda(0)|} \; - \; \log \left| \frac{|f_-(\lambda;x)|}{|f_-(\lambda;0)|} \right| \right| = 0 \qquad \text{P–a.e..}$$

For the proof we write $g_\lambda = a \cdot f_-(\lambda) + b \cdot f_+(\lambda)$ with $a,b \in \mathbf{C}$. The coefficient a must be different from zero because otherwise $\mathrm{Im}[g_\lambda,\bar{g}_\lambda](0) = -2|b|^2 \, |f_+(\lambda;0)|^2 \cdot \mathrm{Im}\, m_+(\lambda) < 0$ in contradiction to assumption (6.57). Since $|f_+(\lambda;x)|$ is exponentially decreasing as $x \to +\infty$ and $|f_-(\lambda;x)|$ is exponentially increasing as $x \to +\infty$ P – a.e. (cf. Figure 12), we get the desired conclusion.

Step 3. Proof of the equality

60) $$E[m_+(\lambda)] = E[m_-(\lambda)].$$

We use the same method as before in step 2 in section 6.2. and try to write $m_+(\lambda;\theta_x q) - m_-(\lambda;\theta_x q)$ in divergence form

1) $$m_+(\lambda;\theta_x q) - m_-(\lambda;\theta_x q) \; + \; \frac{d}{dx} \, H(\theta_x q) = 0 \qquad , x \in \mathbf{R},$$

with a suitable function $H\colon \Omega \to \mathbf{R}$. In order to find H, we transform $(m_+(\lambda;\theta_x q) - m_-(\lambda;\theta_x q))$ by means of the Riccati equations (6.27) as follows:

$$m_+(\lambda;\theta_x q) - \; m_-(\lambda;\theta_x q) =$$

$$= \frac{(m_+(\lambda;\theta_x q))^2 - (m_-(\lambda;\theta_x q))^2}{m_+(\lambda;\theta_x q) + m_-(\lambda;\theta_x q)}$$

$$= - \frac{\frac{d}{dx}\big(-m_+(\lambda;\theta_x q) - m_-(\lambda;\theta_x q)\big)}{-m_+(\lambda;\theta_x q) - m_-(\lambda;\theta_x q)}$$

$$= -\frac{d}{dx}\log(-m_+(\lambda;\theta_x q) - m_-(\lambda;\theta_x q))$$

$$= \frac{d}{dx}\log\left(-\frac{1}{m_+(\lambda;\theta_x q) + m_-(\lambda;\theta_x q)}\right)$$

$$= \frac{d}{dx}\log G_\lambda(0,0;\theta_x q),$$

where (6.39) is used in the last line. This shows (6.61) with $H(q) = -\log G_\lambda(0,0;q)$ and concludes the proof of Proposition 4.

d. Hadamard's factorization theorem and Thouless formula

Let $w(\lambda)$, $\lambda \in \mathbb{C}_+$, be defined as in Proposition 4. We will pass to the limit $\operatorname{Im}\lambda \downarrow 0$ and derive (6.32) and (6.33). The main result is formulated in Proposition 5 below. Before we prove it rigorously, we first argue on a heuristic level and deduce a relation between w and N, which turns out to be basic for the following.

For fixed $L > 0$ the function $\lambda \mapsto \psi_\lambda(L)$ is entire and essentially determined by its zeroes, i.e. by the eigenvalues of the operator $H^{(0,L)}$. Choosing $g_\lambda = \psi_\lambda$ in (6.58) and passing formally to the limit $L \to \infty$, a relation between w and the density of the eigenvalues can therefore be expected.

To substantiate this rough idea, we begin with a zero potential. In this case we use the notation ψ^0_λ, w_0, N_0, γ_0 and G^0_λ and write $\overset{\circ}{\lambda}{}^{(L)}_j = \frac{\pi^2 j^2}{L^2}$ $(j \geq 1)$ for the Dirichlet eigenvalues of the vibrating string of the length $L > 0$. Since

$$\sin z = z \cdot \prod_{j\geq 1}\left(1 - \frac{z^2}{\pi^2 j^2}\right), \quad z \in \mathbb{C}$$

(an introduction to the ideas, which originally led Euler to this formula, can be found in Chapter II,6 in Pólya (1954)), $\lambda \mapsto \psi^0_\lambda(L)$ factorizes and can be written in the form

(6.62)
$$\psi^0_\lambda(L) = \frac{\sin(\sqrt{\lambda}\,L)}{\sqrt{\lambda}} = L \cdot \prod_{j\geq 1}\left(1 - \frac{\lambda}{\overset{\circ}{\lambda}{}^{(L)}_j}\right), \quad \lambda \in \mathbb{C},\ L > 0.$$

We argue heuristically and do not worry about condition (6.57) which is not satisfied for ψ_λ^0. Taking logarithm on both sides of (6.62) and passing formally to the limit $L \to \infty$ we obtain

3)

$$w_0(\lambda) = -\lim_{L\to\infty} \frac{1}{L} \log \psi_\lambda^0(L)$$

$$= -\lim_{L\to\infty} \frac{1}{L} \Big(\log L + \sum_{j\geq 1} \log \Big(1 - \frac{\lambda}{\lambda_j^{0(L)}}\Big)\Big)$$

$$= -\int_0^\infty \log \Big(1 - \frac{\lambda}{x}\Big)\, N_0(dx) \quad ,$$

at least for $\lambda \in C_+$. Differentiating with respect to λ and using (6.10) and (6.22), we conclude from (6.63)

4)

$$\frac{d}{d\lambda} w_0(\lambda) = \int_0^\infty \frac{1}{x-\lambda}\, N_0(dx) = E[G_\lambda^0(0,0)] \qquad , \lambda \in C_+ .$$

In the case of a general potential $q \in \Omega$, the function $\lambda \mapsto \psi_\lambda(L)$ $(L > 0$ fixed) is still entire and of order of growth $\frac{1}{2}$ (notice $\lim_{\lambda\to-\infty} \frac{\psi_\lambda(L)}{\psi_\lambda^0(L)} = 1$, and $\psi_\lambda^0(L) = \frac{e^{i\sqrt{\lambda}L} - e^{-i\sqrt{\lambda}L}}{2i\sqrt{\lambda}}$ is of order of growth $\frac{1}{2}$). By an application of Hadamard's factorization theorem we thus obtain similarly as in (6.62)

)

$$\psi_\lambda(L) = c_L \cdot \prod_{j\geq 1} \Big(1 - \frac{\lambda}{\lambda_j^{(L)}}\Big) \qquad , \lambda \in C,\ L > 0,$$

with a constant c_L. To get rid of c_L, it is expedient to normalize (6.65), for example by

)

$$\frac{\psi_\lambda(L)}{\psi_i(L)} = \prod_{j\geq 1} \frac{\lambda_j^{(L)} - \lambda}{\lambda_j^{(L)} - i} \qquad , \lambda \in C,\ L > 0.$$

Taking logarithm on both sides of (6.66) and passing formally to the limit $L \to \infty$, we conclude

$$w(\lambda) - w(i) = -\int_0^\infty \log\frac{x-\lambda}{x-i}\, N(dx) \qquad , \lambda \in \mathbf{C}_+ , \tag{6.67}$$

and by differentiation with respect to λ

$$\frac{d}{d\lambda}\, w(\lambda) = \int_0^\infty \frac{1}{x-\lambda}\, N(dx) = E[G_\lambda(0,0)] \qquad , \lambda \in \mathbf{C}_+ . \tag{6.68}$$

According to this formula, the rate of change of the Floquet exponent is just the averaged Green's function. Taking real parts on both sides of (6.67) and passing formally to the limit $\operatorname{Im} \lambda \downarrow 0$, one obtains a relation between γ and N. To derive it in a form, which is usual in physics literature, we proceed slightly differently, using instead of (6.66) a normalization of ψ_λ by ψ_λ^0 and writing

$$\begin{aligned} \frac{\psi_\lambda(L)}{\psi_\lambda^0(L)} &= \frac{c_L}{L} \prod_{j \ge 1} \frac{\overset{\circ}{\lambda}{}_j^{(L)}}{\lambda_j^{(L)}} \cdot \prod_{j \ge 1} \frac{\lambda_j^{(L)} - \lambda}{\overset{\circ}{\lambda}{}_j^{(L)} - \lambda} \\ &= \text{constant} \cdot \prod_{j \ge 1} \frac{\lambda_j^{(L)} - \lambda}{\overset{\circ}{\lambda}{}_j^{(L)} - \lambda} \\ &= \prod_{j \ge 1} \frac{\lambda_j^{(L)} - \lambda}{\overset{\circ}{\lambda}{}_j^{(L)} - \lambda} . \end{aligned} \tag{6.69}$$

The constant in (6.69) equals 1, since

$$\lim_{\lambda \to -\infty} \frac{\psi_\lambda(L)}{\psi_\lambda^0(L)} = 1 \quad \text{and} \quad \lim_{\lambda \to -\infty} \prod_{j \ge 1} \frac{\lambda_j^{(L)} - \lambda}{\overset{\circ}{\lambda}{}_j^{(L)} - \lambda} = 1$$

because of $\lambda_j^{(L)} = \overset{\circ}{\lambda}{}_j^{(L)} + O(1)$ (cf. for example Chapter 2 in Pöschel and Trubowitz (1987)). Similarly as before, (6.69) suggests

0)
$$w(\lambda) - w_0(\lambda) = -\int_0^\infty \log(x-\lambda)\,(N(dx) - N_0(dx)) \qquad , \lambda \in \mathbf{C}_+ .$$

Taking real parts on both sides of (6.70), passing to the limit Im $\lambda \downarrow 0$ and using (6.32), we formally arrive at the so called Thouless formula

1)
$$\gamma(\xi) = \gamma_0(\xi) + \int_0^\infty \log|x-\xi|\,(N(dx) - N_0(dx)) \qquad , \xi \in \mathbf{R},$$

with $\gamma_0(\xi) = (\max\{0,-\xi\})^{1/2}$ and $N_0(x) = \frac{1}{\pi}(\max\{0,x\})^{1/2}$.

After these heuristic considerations we give a rigorous proof for (6.68) and deduce some consequences of this fundamental formula.

Proposition 5.

2)
$$\frac{d}{d\lambda} w(\lambda) = E[G_\lambda(0,0)] \qquad , \lambda \in \mathbf{C}_+,$$

and

3)
$$w(\xi + i0) = -\gamma(\xi) + i\alpha(\xi) \qquad , \xi \in \mathbf{R}.$$

Corollary. *The Ljapunov exponent and the integrated density of states are related by*

4)
$$\gamma(\xi) = \text{constant} + \int_0^\infty \log\left|\frac{x-\xi}{x-i}\right| N(dx) \qquad , \xi \in \mathbf{R}.$$

Proof. To prove the Proposition and its Corollary, we proceed in five steps.

Step 1. Proof of (6.72).

By (6.59) and (6.39) we can write

$$w(\lambda) = \frac{1}{2} E[m_-(\lambda) + m_+(\lambda)] = -\frac{1}{2} E[\frac{1}{G_\lambda(0,0)}].$$

We therefore have to show

$$E[G_\lambda(0,0;\theta_x q)] + \frac{\partial}{\partial\lambda} \frac{1}{2} E[\frac{1}{G_\lambda(0,0;\theta_x q)}] = 0. \qquad (6.75)$$

We use the same method as previously in the proof of (6.60), express (6.75) in terms of $m_\pm(\lambda)$ and try to find a function $H: \Omega \to \mathbf{R}$ so that

$$-\frac{1}{m_+(\lambda,\theta_x q) + m_-(\lambda,\theta_x q)} - \frac{\partial}{\partial\lambda} \frac{1}{2} (m_+(\lambda;\theta_x q) + m_-(\lambda;\theta_x q)) + \frac{d}{dx} H(\theta_x q) = 0 \qquad (6.76)$$

for $x \in \mathbf{R}$. (6.76) is indeed true with

$$H(q) = -\frac{1}{m_+(\lambda,q) + m_-(\lambda,q)} \cdot \frac{1}{2} \frac{\partial}{\partial\lambda} (m_+(\lambda;q) - m_-(\lambda;q)).$$

To verify this, we denote the derivative $\frac{d}{dx} m_\pm(\lambda;\theta_x q)$ by $m_\pm'(\lambda;\theta_x q)$, drop the variables and compute

$$\frac{d}{dx} H(\theta_x q) = (-\frac{1}{m_- + m_+} \cdot \frac{1}{2} \frac{\partial}{\partial\lambda} (m_+ - m_-))'$$

$$= \frac{\frac{1}{2}(m_-' + m_+')(\frac{\partial}{\partial\lambda} m_+ - \frac{\partial}{\partial\lambda} m_-)}{(m_- + m_+)^2} - \frac{\frac{1}{2}(\frac{\partial}{\partial\lambda} m_+' - \frac{\partial}{\partial\lambda} m_-')}{m_- + m_+}$$

$$= \frac{\frac{1}{2}(m_-^2 - m_+^2)\left(\frac{\partial}{\partial\lambda} m_+ - \frac{\partial}{\partial\lambda} m_-\right)}{(m_- + m_+)^2} + \frac{1 + m_+ \cdot \frac{\partial}{\partial\lambda} m_+ + m_- \cdot \frac{\partial}{\partial\lambda} m_-}{m_- + m_+}$$

$$= \frac{1}{m_- + m_+} + \frac{1}{2}\frac{\partial}{\partial\lambda}(m_+ + m_-) \, .$$

Step 2. Representation of Rew and of Im w by means of N.

Integrating (6.72) we obtain

7)
$$w(\lambda) - w(i) = -\int_0^\infty \log \frac{x-\lambda}{x-i} \, N(dx) \qquad , \lambda \in \mathbf{C}_+ \, ,$$

as was heuristically anticipated in (6.67). The existence of the integral on the right hand side of (6.77) and of similar integrals will be verified in step 5 below, where also the justification is given for manipulations like integration by parts, which we use in steps 2 – 4 without comment. From (6.77) it follows

8)
$$-\operatorname{Rew}(\lambda) = \delta + \int_0^\infty \log \left|\frac{x-\lambda}{x-i}\right| N(dx) \qquad , \lambda \in \mathbf{C}_+ \, ,$$

with a constant δ. Before taking imaginary parts in (6.77), it is expedient to get rid of the logarithm and to integrate by parts. One obtains

$$w(\lambda) - w(i) = -\,i \int_0^\infty \frac{1}{1+x^2} \, N(x)dx + \int_0^\infty \frac{1}{x-\lambda}\,\frac{1+\lambda x}{1+x^2} \, N(x)\,dx$$

and hence

$$\text{(6.79)} \qquad \operatorname{Im} w(\lambda) = \beta + \int_0^\infty \frac{\operatorname{Im}\lambda}{|x-\lambda|^2} N(x)dx \qquad , \lambda \in \mathbf{C}_+ ,$$

with a constant β.

Step 3. Passage to the limit $\operatorname{Im}\lambda \downarrow 0$.
Let $\xi \in \mathbf{R}$ and $\lambda = \xi + i\varepsilon$, $\varepsilon > 0$. By (6.78) one has

$$\text{(6.80)} \qquad -\operatorname{Re} w(\xi + i\varepsilon) = \delta + \int_0^\infty \log\left|\frac{x-\xi-i\varepsilon}{x-\xi}\right| N(dx) + \int_0^\infty \log\left|\frac{x-\xi}{x-i}\right| N(dx)$$

(the integrals are meaningful, see step 5 below). Since

$$\log\left|\frac{x-\xi-i\varepsilon}{x-\xi}\right| = \frac{1}{2}\log\frac{(x-\xi)^2+\varepsilon^2}{(x-\xi)^2}$$

is monotonically decreasing as $\varepsilon \downarrow 0$, it follows

$$\text{(6.81)} \qquad -\operatorname{Re} w(\xi + i0) = \delta + \int_0^\infty \log\left|\frac{x-\xi}{x-i}\right| N(dx) \qquad , \xi \in \mathbf{R}.$$

By (6.79) one has

$$\text{(6.82)} \qquad \operatorname{Im} w(\xi + i\varepsilon) = \beta + \int_0^\infty \frac{\varepsilon}{(x-\xi)^2+\varepsilon^2} N(x)dx$$

$$= \beta + \frac{1}{\varepsilon}\cdot\frac{1}{\pi}\int_0^\infty \frac{\varepsilon^2}{(x-\xi)^2+\varepsilon^2}\,\alpha(x)dx .$$

The measure N(dx) has no atoms, since

3)
$$\int_{|x-\xi|<1} \log \frac{1}{|x-\xi|}\, N(dx) < \infty \qquad , \xi \in \mathbf{R},$$

as is verified in step 5 below. Hence N and thus α are continuous functions, and one obtains from (6.82) in the limit $\varepsilon \downarrow 0$

4)
$$\operatorname{Im} w(\xi + i0) = \beta + \alpha(\xi) \qquad , \xi \in \mathbf{R}.$$

Step 4. Proof of (6.73).

The constant β in (6.84) must be zero, since for $\xi < 0$ the Green's function

$$G_{\xi+i0}(0,0;q) = \int_0^\infty \frac{\rho(dx;q)}{x-\xi}$$

is real and hence the imaginary part of

$$w(\xi + i0) = -\frac{1}{2}\, E\left[\frac{1}{G_{\xi+i0}(0,0)}\right]$$

is vanishing for $\xi < 0$. This proves

5)
$$\operatorname{Im} w(\xi + i0) = \alpha(\xi) \qquad , \xi \in \mathbf{R}.$$

To prove

6)
$$-\operatorname{Re} w(\xi + i0) = \gamma(\xi) \qquad , \xi \in \mathbf{R},$$

we formulate this equality in terms of functions on the complex plane in order to take advantage of complex analysis. We define $h_1 \colon \mathbf{C} \to \mathbf{R}$ and $h_2 \colon \mathbf{C} \to \mathbf{R}$ by

(6.87) $$h_1(z) = \delta + \int_0^\infty \log\left|\frac{x-z}{x-i}\right| N(dx)$$

and

(6.88) $$h_2(z) = \inf_{x>0} \frac{1}{x} E[\log\| Y_z(x;q)\|] .$$

On the real axis, h_1 coincides with the left hand side of (6.86) because of (6.81), and h_2 coincides with the right hand side of (6.86), since

$$\gamma(\xi) = \lim_{x\to\infty} \frac{1}{x} \log \| Y_\xi(x;q)\| = \inf_{x>0} \frac{1}{x} E [\log\|Y_\xi(x;q)\|] \qquad P-\text{a.e.}$$

according to (3.9) and to the subadditive ergodic theorem. We replace equation (6.86) by the seemingly much more general equality

(6.89) $$h_1(z) = h_2(z)$$

for all $z\in \mathbf{C}$, which in fact is easier to prove, since for $z\in \mathbf{C}\setminus\mathbf{R}$ the equality (6.89) can immediately be verified as follows.
For $z\in \mathbf{C}_+$ one has $h_1(z) = -\operatorname{Re}w(z)$ by (6.78), and
$h_2(z) = \lim_{x\to\infty} \frac{1}{x} \log \| Y_z(x;q)\| = |\operatorname{Re}w(z)| = -\operatorname{Re}w(z)$, $z\in \mathbf{C}_+$,
follows because of (cf. Figure 12)

$$|f_\pm(z;x)| \sim \exp(\pm \operatorname{Re}w(z)\cdot x) \;\text{ for } x\to+\infty \qquad P-\text{a.-e..}$$

For $\lambda\in \mathbf{C}_- = \{z\in \mathbf{C}\colon \operatorname{Im} z<0\}$ one can argue similarly, since a Floquet-Weyl theory can be developed analogously to the case $\lambda\in\mathbf{C}_+$. Thus one concludes that h_1 and h_2 coincide at least outside of the real axis, i.e. outside of a set of Lebesgue measure zero. This already implies that h_1 and h_2 are equal everywhere on $\mathbf{C}$, since both functions are *subharmonic* on $\mathbf{C}$, and two subharmonic functions are identical if they are equal outside of a set of Lebesgue measure zero (see p. 270 in Carmona and Lacroix (1990) for information about subharmonic functions). The reason behind the subharmonicity of

h_1 and h_2 is the following. If $F: D \to C$ is analytic in a domain $D \subset C$ and has no zeroes in D, then $\log F(z)$ is also analytic in D and hence $\log |F(z)| = \mathrm{Re} \log F(z)$ is harmonic in D; if F has zeroes in D, Jensen's formula from complex analysis implies, that $\log |F(z)|$ is still subharmonic in D.

Step 5. Technical supplements.

Because of $q \geq 0$ one gets from (6.43)

$$N(\xi) \leq N_0(\xi) = \frac{1}{\pi} (\max\{0,\xi\})^{1/2} \qquad , \xi \in \mathbf{R},$$

hence $\lim\limits_{\xi \downarrow 0} N(\xi) = 0$ and

$$\int_0^\infty \frac{1}{1+\xi} N(d\xi) = \int_0^\infty \frac{N(\xi)}{(1+\xi)^2} d\xi \leq \int_0^\infty \frac{N_0(\xi)}{(1+\xi)^2} d\xi < \infty .$$

Using this estimate one can verify that the integral

$$\int_0^\infty \log \left| \frac{x-\lambda}{x-i} \right| N(dx) \qquad , \lambda \in C_+ ,$$

does exist. By (6.30) one has $-\mathrm{Re}\, w(\lambda) \geq 0$ for $\lambda \in C_+$. Relation (6.78) therefore implies

$$-\int_0^\infty \log \left| \frac{x-\lambda}{x-i} \right| N(dx) \leq \delta < \infty \qquad , \lambda \in C_+,$$

and hence for $\lambda = \xi + i\varepsilon$ $(\xi \in \mathbf{R}, \varepsilon > 0)$

$$-\int_0^\infty \log\left|\frac{x-\xi-i\varepsilon}{x-i}\right| N(dx) = \frac{1}{2}\int_0^\infty \log\frac{x^2+1}{(x-\xi)^2+\varepsilon^2} N(dx) \le \delta < \infty$$

uniformly in $\varepsilon \in (0,1)$. Using monotonous convergence one obtains from this the existence of the integral

$$(6.90) \qquad -\int_0^\infty \log\left|\frac{x-\xi}{x-i}\right| N(dx) \qquad , \xi \in \mathbf{R}.$$

The right hand side of (6.80) is therefore well defined. Moreover, (6.90) implies

$$\int_{|\xi-x|<1} \log\frac{1}{|x-\xi|} N(dx) < \infty \qquad , \xi \in \mathbf{R},$$

hence (6.83) is verified. This concludes the proof of Proposition 5.

6.4 *Proof of the remaining part of Theorem 2*

To finish the proof of Theorem 2, we first verify the heuristic derivation (6.36); then the remaining part of the proof of implication (4.15) will be given. Using Proposition 5 it is easy to justify (6.36) as follows. Let $A \subset \mathbf{R}^1$ be a bounded Borel set with $\gamma(\xi) = -\operatorname{Re}w(\xi + i0) = 0$ for μ-almost all $\xi \in A$. One has to show

$$(6.91) \qquad \lim_{\xi \downarrow 0} \int_A -\frac{\operatorname{Re}w(\xi+i\varepsilon)}{\varepsilon} \mu(d\xi) = \int_A \frac{d}{d\xi}\alpha(\xi)\, \mu(d\xi).$$

It is allowed to interchange the limit and the integral on the left hand side of (6.91). This follows since $-\operatorname{Re}w$ is harmonic on $\mathbf{C}_+$ and can therefore be represented by its boundary values on the real axis as

$$- \operatorname{Re} w(\xi + i\varepsilon) = \frac{1}{\pi} \int_{-\infty}^{+\infty} \frac{\varepsilon}{(x-\xi)^2 + \varepsilon^2} \gamma(x)\, \mu(dx) \qquad , \xi \in \mathbf{R},\ \varepsilon > 0.$$

Hence $-\dfrac{\operatorname{Re} w(\xi + i\varepsilon)}{\varepsilon}$ is monotonically increasing as $\varepsilon \downarrow 0$, and for the proof of (6.91) it suffices to show

2)
$$\lim_{\varepsilon \downarrow 0} \frac{-\operatorname{Re} w(\xi + i\varepsilon) + \operatorname{Re} w(\xi + i0)}{\varepsilon} = \frac{d}{d\xi} \alpha(\xi)$$

for μ-almost all $\xi \in \mathbf{R}$. To prove (6.92) we use the Cauchy-Riemann differential equations for the real and the imaginary part of w. With $\lambda = \xi + i\eta$ $(\eta > 0)$ we write

$$- w(\lambda) = u(\xi,\eta) + iv(\xi,\eta) ,$$

hence

$$\frac{d}{d\lambda} w(\lambda) = - u_\xi(\xi,\eta) + iu_\eta(\xi,\eta),$$

and obtain

3)
$$\lim_{\varepsilon \downarrow 0} \frac{-\operatorname{Re} w(\xi + i\varepsilon) + \operatorname{Re} w(\xi + i0)}{\varepsilon}$$

$$= \lim_{\varepsilon \downarrow 0} \frac{u(\xi,\varepsilon) - u(\xi,0^+)}{\varepsilon}$$

$$= u_\eta(\xi,0^+)$$

$$= \lim_{\varepsilon \downarrow 0} \operatorname{Im} \left(\frac{d}{d\lambda} w\right)(\xi + i\varepsilon) .$$

The limit in the last line does exist for μ-almost all $\xi \in \mathbf{R}$; this can be seen by means of (6.72) and (6.23) as follows:

$$\text{(6.94)} \qquad \lim_{\varepsilon \downarrow 0} \operatorname{Im}(\tfrac{d}{d\lambda} w)(\xi + i\varepsilon) = \lim_{\varepsilon \downarrow 0} \operatorname{Im} \int_0^\infty \frac{1}{x - (\xi + i\varepsilon)} N(dx)$$

$$= \lim_{\varepsilon \downarrow 0} \frac{1}{\pi} \operatorname{Im} \int_0^\infty \frac{1}{x - (\xi + i\varepsilon)} \alpha(dx)$$

$$= \frac{d}{d\xi} \alpha(\xi) \qquad \mu\text{– a.e..}$$

This proves (6.92).

Finally we show the remaining part of (4.15). Let $A \subset \mathbf{R}^1$ be a bounded Borel set such that

$$\text{(6.95)} \qquad \mu(A) > 0 \quad \text{and} \quad \gamma(\xi) = 0 \text{ for } \mu\text{-almost all } \xi \in A.$$

It has to be proved that under assumption (6.95) the potential is necessarily deterministic in the sense, that $q_{|\mathbf{R}_+}$ can be recovered if the values of q on the negative half axis are known. Because of the stationarity of the potential, one can conclude from this, that q is then deterministic in the sense of the definition given previously in section 4.3. This *inverse* problem is the profoundest part of Theorem 2. The plan of the proof is the following.

$$\{q(x), x \le 0\} \overset{(i)}{\longrightarrow} \{m_-(\lambda), \lambda \in C_+\} \overset{(ii)}{\longrightarrow} \{m_+(\lambda), \lambda \in C_+\} \overset{(iii)}{\longrightarrow} \{q(x), x \ge 0\} .$$

Part (i) is obvious, since $m_-(\lambda)$, $\lambda \in C_+$, is determined by the values of the potential on the negative half axis, as the construction of the Titchmarsh-Weyl functions shows.

Part (iii) results from inverse scattering theory, according to which the potential can be recovered, if the Titchmarsh-Weyl functions are given for $\lambda \in C_+$. For the inverse theory we refer to Marchenko (1986).

Here we concentrate on the proof of (ii). It is based on the theory developed in the previous section. The function $m_-(\lambda;q)$ describes, roughly speaking, the "process of repeated scattering on the individual potential q" in the left half space $\mathbf{R}_-$, and $m_+(\lambda;q)$ describes the "process of repeated scattering on the potential q" in the right half space $\mathbf{R}_+$. As was seen in previous sections, the scattering processes on the left and on the right give the same results *in the average* (see for example (6.59). This is of course not true for individual potentials. However, in view of the Bloch representation (2.16) one expects that, under assumption (6.95), $m_-(\xi + i0;q)$ and $m_+(\xi + i0;q)$ ($\xi \in A$) are related to each other, and one can hope to recover $\{m_+(\lambda), \lambda \in \mathbf{C}_+\}$ from $\{m_-(\lambda), \lambda \in \mathbf{C}_+\}$ by means of such a relation. This is indeed the case as will be shown now.

Lemma. *Assume* (6.95) *Then*

6) $$m_+(\xi + i0;q) = -\overline{m_-(\xi + i0;q)}$$

for $\mu \otimes P$-*almost all* $(\xi,q) \in A \times \Omega$.

Given the Lemma, part (ii) follows immediately, since a holomorphic function $h: \mathbf{C}_+ \to \mathbf{C}_+$ is determined not only by its boundary values on the real axis, but already, if its boundary values on a set of positive Lebesgue measure are known (see for example § 7 in Kotani (1987)). According to the Lemma, the boundary values $m_+(\xi + i0;q)$ are known for μ-almost all $\xi \in A$, if $\{m_-(\lambda;q), \lambda \in \mathbf{C}_+\}$ is given. Because of $\mu(A) > 0$ this proves (ii).

Proof of the Lemma. We try to extract from (6.93) as much information about the boundary values of $m_\pm$ as possible. To do so we use those relations, on which the previously developed theory was mainly based, namely (6.30) and (6.72), and the symmetry $E[m_-(\lambda)] = E[m_+(\lambda)]$ (every of these relations has to do with some kind of conservation property, as their proof by means of the divergence trick shows). In this way we express (6.93) in terms of $m_-(\lambda)$ and of $m_+(\lambda)$ and obtain for $\lambda \in \mathbf{C}_+$

7) $$-\frac{\operatorname{Re} w(\lambda)}{\operatorname{Im} \lambda} - \operatorname{Im}\left(\frac{d}{d\lambda} w\right)(\lambda)$$

$$= \frac{1}{4} E\left[\frac{1}{\operatorname{Im} m_-(\lambda)} + \frac{1}{\operatorname{Im} m_+(\lambda)}\right] + \operatorname{Im} E\left[\frac{1}{m_-(\lambda) + m_+(\lambda)}\right]$$

$$= E[F(m_-(\lambda), m_+(\lambda)]$$

with

(6.98) $F(m_-(\lambda), m_+(\lambda)) =$

$$\frac{1}{4}\Big(\frac{1}{\mathrm{Im}\, m_-(\lambda)} + \frac{1}{\mathrm{Im}\, m_+(\lambda)}\Big) \cdot \frac{(\mathrm{Re}(m_-(\lambda) + m_+(\lambda))^2 + (\mathrm{Im}(m_-(\lambda) - m_+(\lambda)))^2}{|m_-(\lambda) + m_+(\lambda)|^2}$$

≥ 0 .

Since $m_\pm$ is holomorphic on $\mathbf{C}_+$ and has a positive imaginary part, the limit $m_\pm(\xi + i0;q) = \lim_{\varepsilon\downarrow 0} m_\pm(\xi + i\varepsilon;q)$ exists for μ-almost all ξ (see §7 in Kotani (1987)). Moreover, $\mathrm{Im}\, m_\pm(\xi + i0;q) > 0$ for $\mu\otimes P$-almost all $(\xi,q)\in A \times \Omega$ because of (6.37). Inserting (6.97) in (6.93) and using (6.95), (6.98) and Fatou's lemma, we get

$$0 \leq \int_A E[F(m_-(\xi + i0), m_+(\xi + i0))]\, \mu(d\xi)$$

$$= \int_A E[\lim_{\varepsilon\downarrow 0} F(m_-(\xi + i\varepsilon), m_+(\xi + i\varepsilon))]\, \mu(d\xi)$$

$$\leq \int_A \liminf_{\varepsilon\downarrow 0} E[F(m_-(\xi + i\varepsilon), m_+(\xi + i\varepsilon))]\, \mu(d\xi) = 0,$$

hence

$$F(m_-(\xi + i0;q), m_+(\xi + i0;q)) = 0$$

for $\mu\otimes P$-almost all $(\xi,q)\in A \times \Omega$. By (6.98) this proves the Lemma.

Conclusion

Here, we recapitulate the contents of the previous sections and reflect upon the evolution of the ideas which have led to Theorem 1 and to Theorem 2.

Different roles of probability and notions of instability

The notion of probability can play quite different roles in the theory of random Schrödinger operators depending on the kind of questions one is interested in. If one deals with questions motivated by physical problems one is working with special models. Probability is introduced according to the degree of disorder in these models, containing stronger or weaker correlations. On the other hand, for example in Theorem 2, probability is introduced as a tool for averaging in order to provide results which would be out of reach for individual potentials.

As a first step, the spectral problems dealt with in Theorem 1 and in Theorem 2 are both translated into dynamical questions; in the first case via the Feynman-Kac formula into a problem concerning the long time behaviour of Brownian motion, and in the second case, using the one-dimensionality of space, into a question concerning the stability of a deterministic dynamical system of first order. The key notions which allow to analyze stability, are the notion of a generalized entropy in the case of Theorem 1, which is, roughly speaking, defined by

$$I(\rho) = - \lim_{t\to\infty} \frac{1}{t} \log Q\left(L_t \approx \rho\right) \qquad , \rho \in \mathcal{M}(T) ,$$

and the notion of a generalized Floquet exponent in the case of Theorem 2, which is given by

$$w(\lambda) = - \lim_{L\to\infty} \frac{1}{L} \log g_\lambda(L) \qquad , \lambda \in \mathbf{C}_+ .$$

In the first case, the mathematical reasoning is based on inequalities and convex analysis (cf. section 5.3.), in the second case, it is based on analytic identities and complex analysis (cf. section 6.3.). Beautiful introductions to the ideas underlying (7.1) and (7.2) respectively can be found in the Fermi-lectures by Kac (1980) and by Moser (1981). Given these basic concepts, the functional I is used in order to compute probabilities of large deviations for Brownian motion, and the function w and its boundary behaviour on the real axis are used in order to characterize stability of solutions of equation (1.3) in the one-dimensional case.

7.2. *Extensions of classical notions and results*

The basic step in the proofs of Theorem 1 and Theorem 2 consists of an extension of classical notions of Boltzmann and of Floquet respectively. In order to find an extension of Boltzmann's counting method, one starts with a product measure as reference measure and develops a scheme of proof which can be applied to the case of much more general reference measures, e.g. to the case of Brownian motion. The same line of reasoning can be carried further and a notion of entropy can be introduced for stationary processes, see Varadhan (1984). It turns out that the functional I is related to this entropy by

$$I(\rho) = \inf\{H(R|Q^{\mu}): R \text{ is a stationary process on } T \text{ with marginal distribution } \rho\},$$

where Q^{μ} denotes the measure of Brownian motion on T whose starting point is distributed according to the normalized Lebesgue measure on T and where $H(R|Q^{\mu})$ is the entropy of the process R relative to Q^{μ}.

In turn, the extended theory of entropy sheds new light on variational principles of classical analysis like the Dirichlet principle and the Rayleigh-Ritz principle, revealing their relation to the Gibbs variational principle of Statistical Mechanics.

In section 6.2. we have tried to show the way from classical stability theory for Hill's equation to the result formulated in Theorem 2. The key notions were introduced by Johnson and Moser (1982) in the case of almost periodic potentials; we have omitted this intermediate step in our discussion. It turns out that the generalized Floquet exponent is just the average of the Titchmarsh-Weyl functions. By means of Sturm's oscillation theory it can be seen that the imaginary part of $w(\lambda)$ yields, in the limit as λ approaches the real axis, the so called rotation number. The unity of these classical concepts is brought to light with the help of the stationarity of the underlying potential. The proof of the existence of an absolutely continuous spectrum is based on the notion of the rotation number, which plays the role of an analogue to the phase shift in scattering theory. Its archetype can be found in Poincaré's work on one of the simplest dynamical systems with compact state space, diffeomorphisms of the unit circle. For a monotonically increasing differentiable map $f: \mathbf{R} \to \mathbf{R}$ satisfying $f(x + 1) = f(x) + 1$, the rotation number is defined as

$$\alpha = \alpha(f) = \lim_{n \to \infty} \frac{f^n(x)}{n} ,$$

where f^n denotes the n-fold iterate of f.

From the vibrating string to infinitely many randomly coupled membranes and to non-linear wave equations

John Bernoulli was amongst the first to begin the study of the vibrating string. In 1728 he introduced the wave equation in discretized form. On the other hand, one of the first books in the field of probability theory was written by John's brother James Bernoulli. His "Ars conjectandi", in which, among other things, the binomial distribution is introduced and the weak law of large numbers is proved, appeared posthumously in 1713. From these beginnings the ideas evolved in both fields, spectral theory and probability theory, steadily but separately.

In our century, along with the introduction of the Wiener integral, connections between spectral problems and points of view coming from probability theory gradually emerged. As with probabilistic potential theory, the introduction of randomness has given new insights into many branches of the spectral theory of Schrödinger operators. The contents of sections 5 and 6 for example may be viewed as a natural continuation of classical ideas of Courant and Weyl, which are explained in the expository article by H. Weyl (1950).

We started with example 1, which is essentially the case of one vibrating string. In generalizing this example to higher dimensions, one has to deal with a system of infinitely many membranes which are coupled randomly in a very complicated way.

The problems, explained in section 4.1. for example, are genuinely infinite-dimensional and, compared with classical spectral theory, therefore on a qualitatively new level, exhibiting all the difficulties which are typical for problems with phase transitions in Statistical Mechanics. To this increased degree of complication corresponds a narrowness of the range of the results so far obtained. Theorem 1 for example is merely of an asymptotic character, and results of the type (4.2) can only be proved for special models and asymptotically for large disorder.

Whereas equation (1.3) itself is a linear equation, genuinely non-linear problems, which are beyond classical spectral theory, are in the background. This becomes clear if one looks at the problem (1.2) from the left hand side of Figure 1.

The main impetus for trying to extend example 2 to Schrödinger operators with almost periodic potentials came from the study of the Korteweg de Vries (KdV) equation

(7.3)
$$\begin{aligned} \frac{\partial}{\partial t} u(t,x) &= 6 \cdot u(t,x) \cdot \frac{\partial}{\partial x} u(t,x) - \frac{\partial^3}{\partial x^3} u(t,x) \\ u(0,x) &= u_0(x) \quad , \quad t \geq 0 \; , -\infty < x < +\infty \; , \end{aligned}$$

which is non-linear and which can be interpreted, in a certain sense, as an infinite-dimensional Hamiltonian system, see Moser (1981). In the sixties, the following basic relation between the KdV-flow and the spectral theory of one-dimensional Schrödinger operators was discovered. If one denotes by

$$H_t = -\frac{d^2}{dx^2} + q_t$$

the Schrödinger operator whose potential $q_t = u(t, \cdot)$ is given by a solution to the KdV-equation (7.3), $t \geq 0$ playing the role of a parameter, and if the initial condition $u_0 = q_0$ belongs to a suitable class of functions, then the spectrum

$$\Sigma(q_t) \equiv \Sigma(q_0) \quad , \quad t \geq 0,$$

is invariant under the KdV-flow. The following inverse problem is, therefore, fundamental: given the spectrum Σ, find all potentials q such that $\Sigma(q) = \Sigma$ and determine the KdV-flow in this class. In a first step one studies this problem in the case of periodic potentials and then proceeds further to the case of almost periodic potentials, see Moser (1981). Here the connection with the Floquet exponent $w(\lambda) = w(\lambda;q)$, which can be defined pointwise for individual q in the class of almost periodic potentials, appears. If $q_0 = u_0$ is almost periodic in space, then, as it turns out, not only the spectrum $\Sigma(q_t)$ is invariant under the KdV-flow, but also the Floquet exponent $w(\lambda;q_t) \equiv w(\lambda;q_0)$, $t \geq 0$ $(\lambda \in \mathbf{C}_+)$. Moreover, the family $\{w_k(q_0),\ k \in \mathbf{N}\}$ of the coefficients in the asymptotic expansion of $w(\lambda;q_0)$ as $\lambda \to -\infty$ is just the infinite family of conservation laws of the KdV-equation. Under the one roof of problem (1.2) we therefore have the two extremes: on one side the notion of entropy, which is a measure of disorder, and on the other side infinitely many conservation laws indicating the complete integrability of the Hamiltonian system which corresponds to the KdV-flow.

The geometric structure which corresponds to a periodic potential in Figure 1 is a lattice (perfect crystal), whereas that which corresponds to a potential of Poisson type is a completely disordered system of points. Are there geometric structures between these two extremes, e.g. corresponding to almost periodic potentials? The answer is yes. For example, the vertices of a Penrose tiling, whose construction and meaning is explained by Penrose (1986), provide such a structure, see Figure 13 on the next page. It is assembled in a necessarily aperiodic way by tiles which are congruent to one of finitely many types. Similar structures, the so called quasi-crystals, can be produced in the laboratory.

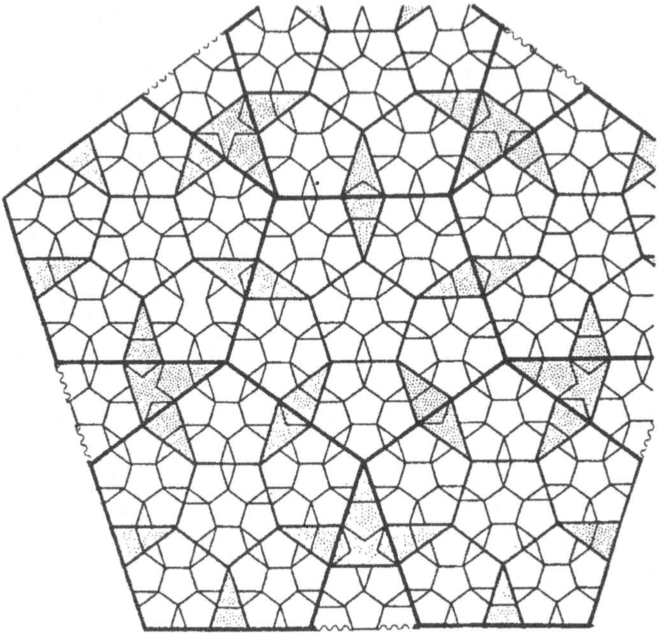

Figure 13

erences

. Anderson (1958). *Absence of diffusion in certain random lattices.* Phys. Rev. **109**, pp. 1492 – 1501.

ubry and P.Y. Le Daeron (1983). *The discrete Frenkel-Kontorova model and its extensions I.* Phys. D **8**, pp. 381 - 422.

ellissard (1989). *Almost Periodicity in Solid State Physics and C^* Algebras.* In: The Harald Bohr Centenary. Proceedings of a Symposium held in Copenhagen 1987, pp. 35 – 75. Edited by C. Berg and B. Fuglede. Copenhagen.

loch and G. Pólya (1932). *On the roots of certain algebraic equations.* Proc. London Math. Soc. **33**, pp. 102 – 114.

armona and J. Lacroix (1990). *Spectral Theory of Random Schrödinger Operators.* Birkhäuser Verlag, Basel, Boston, Berlin.

ourant and D. Hilbert (1953). *Methods of Mathematical Physics.* Volume I. Interscience Publishers, New York.

. Cycon, R.G. Froese, W. Kirsch and B. Simon (1987). *Schrödinger Operators with Application to Quantum Mechanics and Global Geometry.* Springer, Berlin.

. Donsker and S.R.S. Varadhan (1975a). *Asymptotic evaluation of certain Wiener integrals for large time.* In: Functional Integration and Its Applications, pp. 15 - 33. Proceedings of the International Conference Held at Cumberland Lodge, Windsor Great Park, London, April 1974. Edited by A.M. Arthurs. Clarendon, Oxford.

. Donsker and S.R.S. Varadhan (1975b). *Asymptotic evaluation of certain Markov process expectations for large time II.* Comm. Pure Appl. Math. **28**, pp. 279 – 301.

. Donsker and S.R.S. Varadhan (1975c). *Asymptotics for the Wiener sausage.* Comm. Pure Appl. Math. **28**, pp. 525 – 565. Errata, p. 677.

öhlich, T. Spencer and P. Wittwer (1990). *Localization for a Class of One Dimensional Quasi–Periodic Schrödinger Operators.* Commun. Math. Phys. **132**, pp. 5 – 25.

ukushima (1974). *On the spectral distribution of a disordered system and the range of a random walk.* Osaka J. Math. **11**, pp. 73 – 85.

shii (1973). *Localization of eigenstates and transport phenomena in one-dimensional disordered systems.* Suppl. Prog. Theor. Phys. **53**, pp. 77 – 138.

ohnson and J. Moser (1982). *The Rotation Number for Almost Periodic Potentials.* Commun. Math. Phys. **84**, pp. 403 – 438. Erratum, Commun. Math. Phys. **90**, pp. 317 – 318 (1983).

ac (1951). *On some connections between probability theory and differential and integral equations.* Proceedings 2nd Berkeley Symp. Math. Stat. Prob., pp. 189 – 215. Edited by J. Neyman. Univ. of California Press.

ac (1959). *Probability and Related Topics in Physical Sciences.* Interscience, New York.

ac and J. M. Luttinger (1974). *Bose-Einstein condensation in the presence of impurities.* J. Math. Phys. **15**, pp. 183 – 186.

M. Kac (1980). *Integration in Function Spaces. Fermi Lectures*. Academia Nazionale dei Lincei Scuola Normale Superiore, Pisa.

W. Kirsch, S. Kotani and B. Simon (1985). *Absence of absolutely continuous spectrum for some one-dimensional random but deterministic Schrödinger operators*. Ann. Inst. H. Poincaré **42**, 383 – 406.

S. Kotani (1984). *Ljapunov indices determine absolutely continuous spectra of stationary random one-dimensional Schrödinger operators*. In: Proc. Taniguchi Symp. Stochastic Analysis, Katata (1982), pp. 225 – 247. Edited by K. Ito. North Holland, Amsterdam.

S. Kotani (1986). *Lyapunov exponents and spectra for one-dimensional random Schrödinger operators*. Contemporary Mathematics **50**, pp. 277 - 286.

S. Kotani (1987). *One-dimensional random Schrödinger operators and Herglotz functions*. In: Proc. Taniguchi Symp. Prob. Methods in Mathematical Physics, Katata (1985), pp. 219 – 250. Edited by K. Ito and N. Ikeda. North Holland, Amsterdam.

S. Kotani and M. Krishna (1988). *Almost Periodicity of Some Random Potentials*. Journ. Funct. Anal. **78**, 390 – 405.

S. Kotani and B. Simon (1987). *Localization in General One-Dimensional Systems. II. Continuum Schrödinger Operators*. Commun. Math. Phys. **112**, 103 – 119.

I.M. Lifshitz (1965). *Energy spectrum structure and quantum states of disordered condensed systems*. Soviet Physics Uspekhi **7**, 549 – 573.

W. Magnus and S. Winkler (1979). *Hill's Equation*. Dover Publications, New York.

V.A. Marchenko (1986). *Sturm–Liouville operators and applications*. Birkhäuser, Basel, Boston, Stuttgart.

F. Martinelli and E. Scoppola (1987). *Introduction to the Mathematical Theory of Anderson Localization*. Rivista Del Nuovo Cimento **10**, no 10, 1 – 90.

J. Moser (1981). *Integrable Hamiltonian Systems and Spectral Theory*. Fermi Lectures. Academia Nazionale dei Lincei Scuola Normale Superiore, Pisa.

J. Moser and J. Pöschel (1984). *An extension of a result by Dinaburg and Sinai on quasi–periodic potentials*. Comment. Math. Helvetici **59**, 39 - 85.

J. Moser (1986). *Recent developments in the theory of Hamiltonian systems*. SIAM Review **28**, 459 - 485.

L. Pastur (1980). *Spectral properties of disordered systems in one-body approximation*. Commun. Math. Phys. **75**, 179 - 196.

L. Pastur (1989). *Spectral theory of random self-adjoint operators*. Journal of Soviet Mathematics **46**, 1979 - 2021.

R. Penrose (1986). *Hermann Weyl, Space-Time and Conformal Geometry*. In: Hermann Weyl 1885 - 1985, Centenary Lectures delivered by C.N. Yang, R. Penrose, A. Borel at the ETH Zürich, pp. 23 - 52. Edited by K. Chandrasekharan. Springer, Berlin, Heidelberg, New York.

J. Pöschel and E. Trubowitz (1987). *Inverse Spectral Theory*. Academic Press, New York.

Pólya (1954). *Mathematics and plausible reasoning*. Volume I. Princeton University Press, Princeton.

. Richtmyer (1978). *Principles of Advanced Mathematical Physics*. Volume I. Springer, New York.

Ruelle (1979). *Ergodic theory of differentiable dynamical systems*. Publ. Math. IHES **50**, pp. 275 - 306.

Ruelle (1990). *Chaotic Evolution and Strange Attractors*. Lezioni Lincee. Cambridge University Press, Cambridge.

Schrödinger (1946). *Statistical Thermodynamics*. A Course of Seminar Lectures. University Press, Cambridge.

Spencer (1986). *The Schrödinger equation with a random potential*. A mathematical review. In: Proceedings of the 1984 Les Houches Summer School on critical phenomena, random systems, gauge theories, pp. 895 - 942. Edited by K. Osterwalder and R. Stora. North Holland, Amsterdam.

Spitzer (1964). *Electrostatic capacity, heat flow, and Brownian motion*. Z. Wahrscheinlichkeitstheorie Verw. Gebiete **3**, pp. 110 - 121.

. Sznitman (1990). *Lifschitz tail and Wiener sausage I*. J. Funct. Anal. **94**, 223 - 246.

.S. Varadhan (1984). *Large Deviations and Applications*. SIAM, Philadelphia.

Weyl (1950). *Ramifications, old and new, of the eigenvalue problem*. Bull. AMS **56**, pp. 115 - 139.

Index